Selective Breeding in Aquaculture: An Introduction

Reviews: Methods and Technologies in Fish Biology and Fisheries

VOLUME 10

Series editor:

Jennifer L. Nielsen

U.S. Geological Survey,
Alaska Science Center
Anchorage, Alaska

For further volumes:
http://www.springer.com/series/6481

Trygve Gjedrem · Matthew Baranski

Selective Breeding in Aquaculture: An Introduction

Springer

Trygve Gjedrem
Nofima Marin
1432 Aas
Norway
trygve.gjedrem@nofima.no

Matthew Baranski
Nofima Marin
1432 Aas
Norway

ISSN 1571-3075
ISBN 978-94-007-3677-1 ISBN 978-90-481-2773-3 (eBook)
DOI 10.1007/978-90-481-2773-3
Springer Dordrecht Heidelberg London New York

Softcover reprint of the hardcover 1st edition 2009

Cover illustration: Larvae of Atlantic Cod by Saskia Mennen, Nofima

Printed on acid-free paper

Springer is part of Springer Science+Business Media (www.springer.com)

Preface

The foundation of quantitative genetics theory was developed during the last century and facilitated many successful breeding programs for cultivated plants and terrestrial livestock. The results have been almost universally impressive, and today nearly all agricultural production utilises genetically improved seed and animals. The aquaculture industry can learn a great deal from these experiences, because the basic theory behind selective breeding is the same for all species.

The first published selection experiments in aquaculture started in 1920 s to improve disease resistance in fish, but it was not before the 1970 s that the first family based breeding program was initiated for Atlantic salmon in Norway by AKVAFORSK. Unfortunately, the subsequent implementation of selective breeding on a wider scale in aquaculture has been slow, and despite the dramatic gains that have been demonstrated in a number of species, less than 10% of world aquaculture production is currently based on improved stocks. For the long-term sustainability of aquaculture production, there is an urgent need to develop and implement efficient breeding programs for all species under commercial production. The ability for aquaculture to successfully meet the demands of an ever increasing human population, will rely on genetically improved stocks that utilise feed, water and land resources in an efficient way. Technological advances like genome sequences of aquaculture species, and advanced molecular methods means that there are new and exciting prospects for building on these well-established methods into the future.

The main purpose of this book is to demonstrate the success that selective breeding programs have achieved so far in aquaculture, and to highlight the tremendous potential this technology offers for efficient and productive aquaculture production in the future.

The main sections of the book are:

- Why improve production traits in fish and shellfish?
- What has been accomplished in selective breeding programs in aquaculture?
- A brief outline of the theory of quantitative genetics
- Establishing and running breeding programs
- Integration of molecular genetic tools

The book is primarily written for aquaculture students with selective breeding as a subject, farmers, advisory consultants and farm managers. Students specialising in selective breeding may also find it useful to consult the book 'Selection and breeding programs in aquaculture' (Springer, 2005), which provides a more in-depth coverage of the topics discussed here.

We hope that this book will stimulate aquaculture industries to consider the use of improved stocks in their production of fish and shellfish. The development and implementation of breeding programs must be driven by industry, with the support of scientists, farmers organisations and governments. The benefits will be far reaching.

Ås, Norway
February 2009

Trygve Gjedrem
Matthew Baranski

Acknowledgements

We express our sincere appreciation to our employer, Nofima Marin, Ås, for their continuous support and financial contribution to the publication of this book. It has been inspiring to work together with the skilled scientists at the institute. We are particularly grateful to Dr. Solveig van Nes, Reidun Lilleholt and Grethe Tuven for technical help. Furthermore, we thank the director of AKVAFORSK Genetics Center AS (AFGC), Dr. Morten Rye, for his valuable contribution to the manuscript. Finally, we would like to thank Dr. Hans B. Bentsen, Dr. Bjarne Gjerde and Dr. Nicholas Robinson for their valuable support.

Ås, Norway
February 2009

Trygve Gjedrem
Matthew Baranski

Contents

1 **Introduction** 1
   1.1 Historic Development of Aquaculture 1
   1.2 Definition of a Breeding Program 3

2 **Domestication and the Application of Genetic Improvement in Aquaculture** 5
   2.1 Domestication of Animals 5
   2.2 Selective Breeding 6
   2.3 Quality Traits 7
   2.4 Better Utilization of Resources 7
   2.5 Genetic Improvement is Accumulative 7
   2.6 Genetic Improvement Produces Permanent Gains 8
   2.7 Initiating a Selective Breeding Program 9
   2.8 Selective Breeding Programs in Aquaculture 10
   2.9 Prerequisites for a Breeding Program 11

3 **The Success of Selective Breeding in Aquaculture** 13
   3.1 Introduction 13
   3.2 Atlantic Salmon 13
   3.3 Rainbow Trout 15
   3.4 Coho Salmon 16
   3.5 Tilapia 17
   3.6 Carp 18
   3.7 Channel Catfish 19
   3.8 Sea Bream 20
   3.9 Shrimp 20
   3.10 Oysters 21
   3.11 Scallops 21
   3.12 Genetic Improvement in Aquatic Species Compared to Terrestrial Livestock Species 22
   3.13 Summary and Conclusion 22

4 **The Theoretical Basis for Breeding and Selection** 25
   4.1 Introduction 25
   4.2 The Cell 25

4.3 Basic Genetics . . . 26
4.3.1 Introduction . . . 26
4.3.2 Genes . . . 26
4.3.3 Effect of Genes . . . 29
4.4 Variation . . . 29
4.4.1 Introduction . . . 29
4.4.2 Single Gene Traits . . . 30
4.4.3 Quantitative Traits . . . 30
4.4.4 Variation in Quantitative Traits . . . 31
4.4.5 Variation Between Species . . . 32
4.4.6 Variation Within Species . . . 33
4.5 Estimation of Variation and Covariation . . . 33
4.5.1 Mean and Standard Deviation . . . 33
4.5.2 Variance of a Sum . . . 35
4.5.3 Genetic Variance . . . 36
4.5.4 Heritability . . . 36
4.5.5 Environmental Variance . . . 38
4.5.6 Correlations Between Traits . . . 40
4.5.7 Regression . . . 42
4.6 Inbreeding and Relatedness . . . 43
4.6.1 Genetic Relationship . . . 43
4.6.2 Inbreeding . . . 44
4.6.3 Effective Population Size . . . 47
4.6.4 Effect of Inbreeding on Genetic Variance . . . 48
4.6.5 Inbreeding Depression . . . 48
4.7 Crossbreeding . . . 50
4.7.1 Introduction . . . 50
4.7.2 Heterosis . . . 50
4.8 Purebreeding . . . 52
4.9 Selection . . . 52
4.9.1 Introduction . . . 52
4.9.2 Natural Selection . . . 53
4.9.3 Artificial Selection . . . 53
4.9.4 Predicting Selection Change . . . 54
4.9.5 Multiple Trait Selection . . . 56
4.9.6 Correlated Response to Selection . . . 57
4.9.7 Effect of Selection on Genetic Variance . . . 58
4.9.8 Methods of Selection . . . 58
4.9.9 Selection Limits . . . 59

5 Initiating Breeding Programs . . . 63
5.1 Introduction . . . 63
5.2 The Fundamental Basis of a Breeding Program . . . 64
5.3 Establishment of a Base Population . . . 65

5.4 Breeding Goal . . . 67
- 5.4.1 Introduction . . . 67
- 5.4.2 Growth Rate . . . 68
- 5.4.3 Feed Conversion Ratio (FCR) and Efficiency (FCE) . . 69
- 5.4.4 Disease Resistance . . . 70
- 5.4.5 Age at Sexual Maturation . . . 74
- 5.4.6 Product Quality . . . 75
- 5.4.7 Cold Tolerance . . . 78
- 5.4.8 Fecundity . . . 78
- 5.4.9 Behaviour . . . 78
- 5.4.10 Recapture Frequency . . . 79
- 5.4.11 Central Breeding Goals . . . 79

5.5 Registration of Records . . . 80
- 5.5.1 Introduction . . . 80
- 5.5.2 Body Weight . . . 81
- 5.5.3 Survival . . . 81
- 5.5.4 Feed Conversion Efficiency . . . 82
- 5.5.5 Product Quality . . . 82

5.6 Adjustment of Data . . . 83

6 Breeding Strategies . . . 87

6.1 Introduction . . . 87

6.2 Inbreeding . . . 87

6.3 Crossbreeding . . . 88

6.4 Purebreeding . . . 90

7 Selection Methods . . . 93

7.1 Introduction . . . 93

7.2 Factors Affecting Allele Frequencies . . . 93
- 7.2.1 Migration . . . 93
- 7.2.2 Selection . . . 93
- 7.2.3 Mutation . . . 94
- 7.2.4 Genetic Drift . . . 94

7.3 Choice of Selection Method . . . 94

7.4 Pedigree Selection . . . 95

7.5 Individual Selection . . . 95

7.6 Family Selection . . . 96

7.7 Within-Family Selection . . . 98

7.8 Progeny Testing . . . 98

7.9 Correlated Response and Indirect Selection . . . 99

7.10 Combined Selection . . . 100

7.11 Multiple Trait Selection and Index Selection . . . 101

7.12 Comparing Different Selection Methods . . . 102

8 Mating Design . . . 105

8.1 Introduction . . . 105

8.2 Mass Spawning . . . 105

8.3 Single Pair Mating 106
8.4 Nested Mating Design 108
8.5 Factorial Mating 109
8.6 Connectedness 110
8.7 Conclusion 111

9 Estimation of Breeding Values 113
9.1 Introduction 113
9.2 Breeding Value of Individual Animals 114
9.3 Breeding Value of Full-Sib Families 114
9.4 Breeding Value of Half-Sib Families 115
9.5 Breeding Values for Multiple Traits Using a Selection Index . . 115
9.6 Scaling of Selection Indexes 116
9.7 Best Linear Unbiased Prediction (BLUP) 117

10 Genotype–Environment Interaction 119
10.1 Introduction 119
10.2 Estimates of Genotype–Environment Interactions 119
10.3 Conclusion 124

11 Measuring Response to Selection 125
11.1 Introduction 125
11.2 Control Population 126
11.3 Average Breeders 126
11.4 Repeated Matings 128
11.5 Genetic Trend Analysis 129
11.6 Conclusion 129

12 Structure of Breeding Programs 131
12.1 Introduction 131
12.2 Breeding Programs Applying Individual Selection 131
12.3 Advanced Breeding Programs 132
12.3.1 Mating and Hatching 136
12.3.2 Tagging 136
12.3.3 Recording During Grow-Out 137
12.3.4 Harvest and Pre-selection of Broodstock at Breeding Stations 137
12.3.5 Final Selection of Broodstock 138
12.3.6 Genetic Markers for Parentage Assignment 138
12.4 Test Stations 138
12.5 Production of Special Lines 139
12.6 Dissemination of Genetic Gains 140
12.6.1 Direct Dissemination from Nucleus 140
12.6.2 Dissemination from Multipliers 142
12.7 Breeding Programs for New Species 143

13 Undesirable Side Effects in Breeding Programs 145
13.1 Introduction 145
13.2 Correlated Effects 145
13.3 Breeding Goal May Change 147
13.4 Disease Prevention 147
13.5 Genotype–Environment Interaction 148
13.6 Increase of Inbreeding 148
13.7 Conclusion 148

14 Biotechnology in Breeding Programs 151
14.1 Introduction 151
14.1.1 DNA Markers 151
14.1.2 Microsatellites 152
14.1.3 Single Nucleotide Polymorphisms (SNPs) 152
14.2 Linkage Maps 153
14.3 Quantitative Trait Loci (QTL) 155
14.3.1 Candidate Gene Approach 156
14.3.2 QTL Mapping Approach 156
14.3.3 Linkage Analysis 157
14.3.4 Fine Mapping of QTL 157
14.3.5 LDLA Mapping 158
14.3.6 An Example of QTL Mapping to Gene Discovery . . . 158
14.3.7 Strategies to Reduce Genotyping Requirements 159
14.4 Marker Assisted Selection 160
14.4.1 Types of Marker Assisted Selection 160
14.4.2 Gene Assisted Selection (GAS) 161
14.4.3 Linkage Disequilibrium MAS (LD-MAS) 162
14.4.4 Linkage Equilibrium MAS (LE-MAS) 162
14.4.5 Genomic Selection 162
14.5 Other Applications of Genetic Markers 163
14.5.1 Parentage Assignment and Traceability 163
14.5.2 Genetic Interactions 164
14.5.3 Genetic Variation 164
14.6 Gene Expression Data 165
14.7 Transgenics 165
14.8 Genome Sequencing and Future Technologies 167

15 Reproduction Techniques 169
15.1 Introduction 169
15.2 Gynogenesis 169
15.3 Androgenesis 171
15.4 Triploidy 171
15.5 Tetraploidy 173
15.6 Production of Single Sex, YY and XX Stocks 173

16 Economic Benefits of Breeding Programs 175
16.1 Introduction 175
16.2 Cost of Broodstock Production 175
16.3 Cost of Running a Breeding Program 176
16.4 Economic Benefit of Breeding Programs 177
16.5 Relative Contribution of Selection and Feed Regimes to Performance 178
16.6 Who Benefits from Genetic Improvements? 179
16.6.1 The Animal 179
16.6.2 The Farmer 181
16.7 Ownership of Genetically Improved Material 183

Appendix 185

Glossary 187

References 199

Index 217

Chapter 1
Introduction

1.1 Historic Development of Aquaculture

Production of fish under controlled conditions started 4–5,000 years ago in China with the farming of common carp *(Cyprinus carpio)* in earthen ponds. Robust omnivores, these carp survived on algae, plankton, snails and detritus that were naturally produced in the ponds. The first known book describing aquaculture, 'ON PISCICULTURE', was written by Fan Li in ancient China and dates back to 475 B.C.

The world production of fish, crustaceans and molluscs has risen dramatically since the mid 1980s (Fig. 1.1) and continues to increase at a rate of around 8% per year. This represents by far the fastest increase in animal protein production today. The annual catch from world fisheries has now stabilised at around 90 million tons (FAO 2007) of which approximately 60 million tons are used for human consumption, and is not expected to increase in coming years. Annual aquaculture production is now approaching the output of wild fisheries, and if current trends continue, will reach this level in five to seven years.

Asia dominates world aquaculture, accounting for around 90% of world production, followed by Europe with 5%, North and South America with 4%, Africa with 1% and Oceania with 0.3% of production. China is by far the highest producing country with 67% of world production. Outside Asia, Chile and Norway are the largest producers of fish and shellfish.

Production of the major aquatic species has risen by an average of 85% in the last 10 years. The various carp species represent the largest share of total production (38%). Crucian carp, tilapia and shrimp have had the highest production increase over the last 10 years. In terms of total production, the Pacific cupped oyster leads with 4,497 million tons in 2005 (Table 1.1).

Aquaculture production of fish originated in freshwater and today freshwater species continue to represent the majority of overall production (54%), with 6% of fish production occurring in brackish waters and only 4% in marine waters. Marine aquaculture production is dominated by molluscs which represent 28% of total aquaculture production, with crustacean species representing 8% of total aquaculture production. Given that saltwater offers by far the largest available area for

T. Gjedrem, M. Baranski, *Selective Breeding in Aquaculture: An Introduction*, Reviews: Methods and Technologies in Fish Biology and Fisheries 10, DOI 10.1007/978-90-481-2773-3_1,

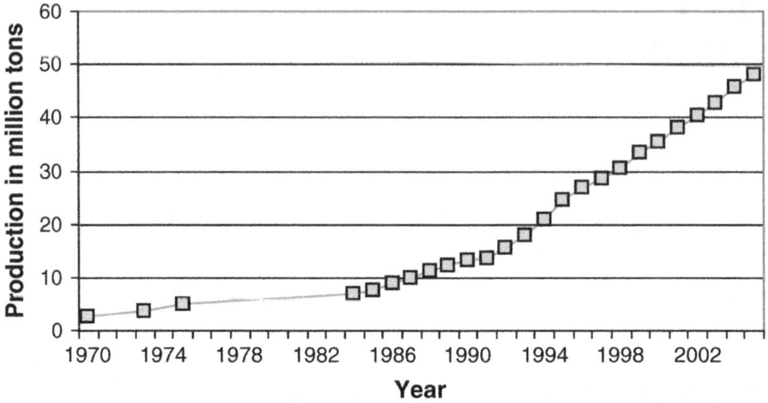

Fig. 1.1 The rise in world aquaculture production

Table 1.1 Production of major aquaculture species in 1996 and 2005 (thousands of tons), (FAO 2007)

Species	1996	2005	Increase %
Common carp	2,041	3,044	49
Grass carp	2,462	3,905	59
Silver carp	2,925	4,153	42
Bighead carp	1,418	2,209	57
Crucian carp	693	2,086	201
Catla	536	1,236	131
Rohu	644	1,196	86
Mrigal	507	330	–35
Nile tilapia	624	1,703	173
Channel catfish	216	380	76
Atlantic salmon	552	1,236	124
Rainbow trout	384	487	27
Milkfish	371	595	60
Yellowtail	146	173	18
Sea bream	113	242	114
Shrimp	917	2,675	192
Pacific cupped oyster	2,925	4,497	54
Japanese carpet shell	1,156	2,947	155
Blue mussel	408	391	−4
Yesso scallop	1,265	1,240	−2
Total	26,592	48,150	85 (avg.)

aquaculture (after all, it covers 72% of the earths surface!), it is anticipated that the largest growth in aquaculture will occur in marine waters in coming years.

Aquaculture development in the future faces many challenges, as an increasing focus is placed upon environmental impacts, competition for water resources intensifies and demand for aquaculture product increases through population growth and declining wild catches. This means that the efficient use of these resources will be of paramount importance for sustainable growth in the sector. In order to achieve this sustainable growth though increasing production efficiency and competitiveness, and enhancing product quality, optimisation of feed and rearing conditions, reduction of production losses, disease management, and genetic improvement are key factors.

1.2 Definition of a Breeding Program

Breeding started with controlled mating and recording of the parents and grandparents of the individuals being bred. This was the beginning of pedigree breeding and development of breeds or strains applying purebreeding practices. In terrestrial livestock, the development of breeds was the main strategy employed from the 1700 s. Breed characteristics were defined and breeding was performed to ensure that these breeds were as uniform as possible in size, body shape, colour and colour pattern. Such a breeding strategy inadvertently resulted in a large degree of inbreeding, and as the breeds became more and more inbred, productivity often decreased. To overcome this problem, it became common to cross different breeds to obtain hybrid vigour.

When the results of Mendel's experiments with plants become known around 1900, the theory of genetics and quantitative genetics was established. The development of the theory of animal breeding was pioneered by Sewall Wright and Jay L Lush. In his book 'Animal breeding plans', first published in 1937, Jay L. Lush discussed the principles and elements of breeding plans for animals.

Lush stressed the importance of making a detailed plan for a breeding program. Such a plan should contain a description of the formation of the founder population, breeding goal, mating strategies, selection procedures and many other factors. The necessity of such a detailed plan may not be obvious immediately, but since results of the breeding work will become apparent in future generations its importance cannot be overstressed.

Chapter 2
Domestication and the Application of Genetic Improvement in Aquaculture

2.1 Domestication of Animals

Price (1984) defines domestication as 'that process by which a population of animals becomes adapted to man and to the captive environment by some combination of genetic changes occurring over generations and environmentally induced developmental events recurring during each generation'. Price (2002) concludes that 'Domestication is about adaptation to man and the environment he provides. Phenotypic adaptations to the captive environment will occur based on the same evolutionary processes that enable free-living populations to adapt to changes in their environment. The major difference is that in captivity, man can accelerate phenotypic changes that would otherwise not appear or persist in nature, through artificial selection'.

Acknowledged as the father of animal breeding, J.L. Lush discusses domestication of farm animals in his pioneering book 'Animal breeding plans' (1949). 'Domestication implies bringing the animals growth and reproduction at least partly under man's control and that man converts the animal's products or services to his own advantage or purposes'. Lush defines domestication primarily in the context of tameness and is usually manifested in rather large changes in behaviour. This is in agreement with Ruzzante (1994) who states that behavioural traits are among the first traits to be affected by the domestication process.

Farm animals have been kept in captivity for thousands of years. During this time the animals have been adapted to the environmental conditions under which they have been reared. The animals tend to become tamer, losing their fear of people and farming enclosures such as fences. In the aquaculture environment under conditions where food is available in excess, domestication can also lead to lower levels of aggression in fish (Doyle and Talbot 1986).

The main process behind domestication is selection. Through the process of natural selection, animals that are best adapted to the particular environment will produce more progeny that survive compared to those that are less adapted. In a controlled farming environment, directional selection will also take place since the farmer will tend to select animals that have the best behaviour and fastest growth. Over time the animals will thrive better, be less nervous, become more resistant to

T. Gjedrem, M. Baranski, *Selective Breeding in Aquaculture: An Introduction*, Reviews: Methods and Technologies in Fish Biology and Fisheries 10,
DOI 10.1007/978-90-481-2773-3_2,

diseases and overall productivity will increase. Simply put, domesticated fish are better adapted to farming conditions (Vandeputte and Prunet 2002).

More recently, signatures of domestication have been identified at a chemical and molecular level. An example of this is the finding of lower cortisol ('the stress hormone') levels in domesticated animals compared with wild animals (Kharlamova and Gulevitch 1991). Pottering (2006) demonstrated that plasma cortisol elevation is a major factor responsible for the damaging effects of stress on survival, growth and reproduction. The influence of stress on growth was experimentally evaluated by Jentoft et al. (2005) who found that stressed rainbow trout and Eurasian perch showed lower body growth than non stressed control fish. In addition to this well documented example of Vandeputte and Prunet (2002) state that domesticated fish are better adapted to farming conditions.

Domestication is a slow process that takes place over a long period of time as animals become increasingly better adapted to their environment. One could hypothesise that domestication would be complete at a particular time point once animals are 'fully' adapted to their particular farming environment. However, the nature of farming is that technology and husbandry practices are constantly improving and evolving, and as a result domestication is more or less a continuous process in many farmed species.

In the case of the earliest cultured aquatic species, carp, it is unclear if domestication commenced as early as the recorded beginning of carp aquaculture (3000–4000 years ago). This uncertainty is primarily due to the fact that the high fecundity of carp meant that it was a relatively simple task to obtain eggs and fry without the need to hold broodstock generation after generation. A number of fish and shellfish species have been under the process of domestication for many generations, however in the last decades a substantial number of new species have been held and cultured in captivity. The domestication of Atlantic salmon, a species with one of the world's highest production levels today, began in Norway and Scotland in the late 1960 s.

Bilio (2007–2008) presented an extensive study of 'Controlled reproduction and domestication in aquaculture – the current state of the art'.

2.2 Selective Breeding

Although the process of domestication alone will result in production improvements and increased efficiency through adaptation to the farming environment, focused selection, or selective breeding, can deliver even more dramatic gains. Selective breeding exploits the substantial genetic variation that is present for the majority of traits with desirable qualities. In the majority of aquaculture species, improvement in growth rate has been the initial focus. As growth rate increases, production time will automatically be reduced along with maintenance requirements, and feed conversion rate will be improved. This means that higher production can be obtained within the technical resources in each farm. The success and gains achieved by selective breeding are described in more detail in Chapter 3.

2.3 Quality Traits

For many aquaculture species, there are distinct consumer preferences for particular quality traits, such as flesh colour, tenderness and flavour. In aquaculture it is possible to produce a product with the quality that the consumer wants through the application of selective breeding, use of optimal tailored feeds and in particular the use of appropriate management practices and culture technologies. For quality traits such as these, the control of the culture environment offers a distinct advantage over wild catch fishery product, where essentially the catch is dictated simply by what is available. Potentially the biggest problem with the improvement of quality traits is not the actual improvement practices themselves, but actually defining the quality that the consumer desires. In fact, even the definition of the traits themselves can be difficult, since they vary between markets. Quality traits are also more prone to changing preferences over time, which necessitates changes in the breeding goal and other farming practices.

2.4 Better Utilization of Resources

When growth rate is improved, it has repeatedly been shown that the animals utilize feeds more efficiently and therefore require less feed to grow a given amount than they did previously. In some cases, feed represents more that half of the production cost, therefore this is of great economic importance. Given that feed resources will become more limited in the future, particularly for wild catch fish meal based diets, such improvement will be critical to ensure continued growth in aquaculture.

Freshwater aquaculture utilises valuable land areas and must frequently compete with other forms of agriculture. The productivity of plant agriculture through selective breeding, extensive use of fertilisers and improved irrigation practices, has dramatically increased and unless freshwater aquaculture production shows similar improvements, there will be a large disparity in productivity of the two culture systems.

In many parts of the world the availability of water itself is a key limiting factor. Water is of course essential for fish and shellfish farming. In many areas of the world, increasing competition for available water will result in less being available for freshwater aquaculture. The only means to maintain or increase production under this scenario is to improve productivity of the animals such that they perform as well or better under the more challenging environmental conditions that they would be subject to.

2.5 Genetic Improvement is Accumulative

Genetic improvement can be illustrated by a stair as is shown in Fig. 2.1. Genetic change through selection can only occur with the production of a new generation.

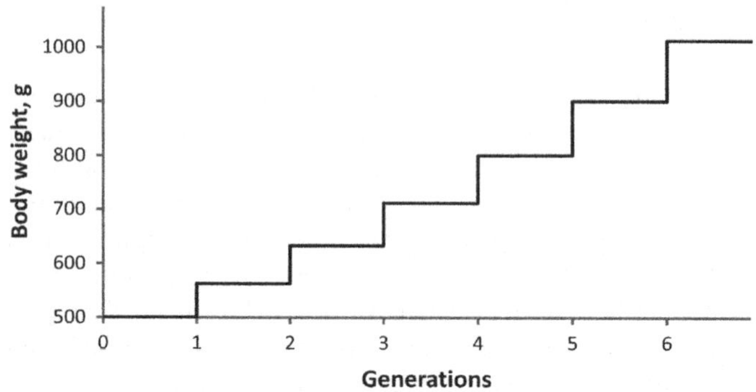

Fig. 2.1 Result of selection for growth rate over six generations when genetic improvement is 12.3% each generation (a figure that has been obtained in several breeding programs). The generation interval is illustrated by the run of the stair (distance between 1st and 2nd) and the rise of the stair is the improvement in body weight as a result of selection

Through selection of parents with favourable phenotypes such as high growth or increased disease resistance, the variants (genes) underlying these traits will be inherited by the progeny; i.e. passed onto the next generation. The generation interval is known as the average time interval between the birth of parents and the birth of their offspring. In Fig. 2.1, the generation interval is illustrated by the run of the stair and the rise of the stair represents the genetic improvement made in each generation through selection.

The stair also illustrates that improvement made in the 2nd generation builds on the improvement already obtained in the 1st generation and that it is not necessary to go back to the starting point or zero generation. This highlights the crucial point that selective breeding is accumulative since the genetic improvement increases continually. The new generation of animals builds on what has been achieved in the parent's generation.

2.6 Genetic Improvement Produces Permanent Gains

A key benefit of selective breeding is that genetic improvement obtained is permanent. In other words, the sum of the genetic improvement made in one generation will be transferred to the following generations.

Compared to terrestrial animals, both males and females of aquatic species are generally very fertile. This facilitates relatively easy dissemination of genetically improved material to the whole industry from a breeding nucleus. Due to the threat of the spread of disease, transport of biological material across country borders is frequently forbidden with a number of exceptions. At present, there is a significant international trade with genetic material (salmon, tilapia and shrimp).

The economic benefit of breeding programs is very high even under the most conservative assumptions and is discussed in Chapter 16.

2.7 Initiating a Selective Breeding Program

Invariably, separate breeding program schemes for different species must be implemented due to variation in fecundity, economic importance of traits and differences in environmental rearing conditions of the species in question. In addition, for health security reasons, each species is usually reared under separate conditions.

Although a broad generalisation, selective breeding programs can be approximately classified into two groups:

- Simple breeding programs applying individual selection, usually for growth rate
- Family based breeding programs selecting for multiple economically important traits.

Individual selection can generally be performed without special technical investments and can be handled by each hatchery/farm. It can only be applied for traits that can be measured or recorded on individual animals which in practice means body weight or length. Usually, fish are not individually tagged. In its most basic form, individual selection is simply a matter of selecting the largest individuals to be used as parents for next generation. The biggest problem with the application of this type of individual selection is the difficulty in avoiding inbreeding. This is especially the case when tagging is not practised as the parentage of selected broodstock is unknown. Individual selection has been widely applied in many aquatic species and has the major advantages of being cheap, easy and doesn't require major infrastructure investments (e.g. separate rearing facilities).

Family based breeding programs require investments in technical equipment because each family must be reared in separate tanks, ponds or hapas from the fertilized egg stage until the fingerlings are large enough to be physically tagged. The needed investments in barns and tanks are relatively high compared with use of hapas.

Usually, two types of families are produced: full-sibs, where all animals in the group have the same father and the same mother, and half-sibs, where two or more families have one parent in common. Due to the level of investment required for separate hatching trays, rearing tanks with automatic feeders, ongoing rearing and multiple trait recording, a family based breeding program generally must serve a number of farms to be economically viable. Large farms covering the whole production cycle may have the economic strength to run their own family based breeding program, but more typically groups of farmers cooperate to initiate such a breeding program.

The first family based breeding program for an aquatic species was started by Institute of Aquaculture Research, AKVAFORSK, in 1973 for rainbow trout and in 1975 for Atlantic salmon. Farmer's organisations or associations would be a natural source to start family based breeding programs for their members. In terrestrial livestock species, cooperative breeding programs have been common.

2.8 Selective Breeding Programs in Aquaculture

The aquaculture industry was late to realise the potential of efficient breeding programs and thus benefit from the use of genetically improved stocks. Early implementation of individual selection for growth rate was employed, but met with varying results. The main reason was that these cases used only few broodstock in each generation because of their high fecundity. Two females provided sufficient eggs and one or two males were sufficient for fertilization, as a result after a few generations of selection the stock became highly inbred. This inbreeding depression resulted in slow growth rate and high mortality. With the failure of individual selection due to rapid inbreeding depression, the immediate alternative was to return to the rivers to obtain wild broodstock once again.

Since the first family based breeding programs for Atlantic salmon and rainbow trout were started in the early 1970 s, several efficient breeding programs have also been started in the 1990 s for other species including tilapia and shrimp. However, the development of new breeding programs in the aquaculture industry has been very slow despite the large genetic improvements that have been demonstrated. In 1992 it was estimated that only 1% of world aquaculture production was based on improved stocks (Gjedrem 1997) and this had only increased to about 5% in 2002 (Table 2.1).

Table 2.1 Impact of selective breeding programs on the production of different aquaculture species

Species	No. programs[1]	No. families per program	World prod. in 2003 (1000 tons)	Prod. From improved stocks (%)
Chinese carps	4	76	15,332	?
Oysters	3	60	4,489	1
Indian carps	1	50	1,796	?
Shrimp	9	170	1,752	8
Tilapia	7	166	1,704	9
Mussel	1	60	1,410	?
Scallop	1	110	1,178	2
Atlantic salmon	12	211	1,129	97
Rainbow trout	7	160	483	27
Channel catfish	1		300	?
Sea bream	1	50	202	?
Pacific salmon	6	108	129	22
Sea bass	1	50	76	?
Crayfish	1	30	14	?
Turbot	1	50	5	?
Arctic charr	1	150	1	?
Atlantic cod	3	90	1	?
Total listed species	60	?	35,051	4.6
Total all species	--	--	42,304	3.8

[1]Number of programs using sib information in the selection decisions
Source: Gjerde et al. 2007a, modified from Gjedrem 2004.

A unique advantage of a breeding program is that one breeding station will be able, if necessary with help of multiplier stations, to disseminate the genetic improvement with minimum delay to all farmers within the country or even broader regions if disease restrictions do not prevent it. As will be discussed later, the benefit/cost ratio of running a selective breeding program is overwhelmingly high.

2.9 Prerequisites for a Breeding Program

Some basic conditions must be met before a breeding program will be efficient:

- There must be variation between animals for the traits under prospect of selection since if all animals share identical phenotypes, there are no individuals with higher than average trait values to select
- A portion of this variation must be due to genetic differences since it is only the genetic variation that is transferred to the next generation through eggs and sperm
- The lifecycle for the species in question must be known and able to be controlled since it must be possible to evaluate progeny for trait characters, subsequently select parents for the next generation and cross them in a controlled manner
- Individual animals must be identifiable (through various tagging methods) in order to keep track of their pedigree.

Chapter 3
The Success of Selective Breeding in Aquaculture

3.1 Introduction

In the previous chapter, it was shown that only a low percentage of aquaculture production across the world is based on genetically improved stocks. One important reason for this slow uptake of the selection technology and development of breeding programs in the aquaculture sector is that good documentation of the genetic gains achieved in such programs has been difficult to obtain. As a result, it has been difficult to quantify the great economic benefits that successful breeding programs produce. Ideally, breeding programs should from the very beginning focus on documentation of realized genetic gains and the economic value of the responses obtained.

However, accurate measures of selection response are difficult to obtain. Documentation of the genetic changes may be important for product marketing, but more importantly, it enables the breeder to make adjustments and fine tune the selection scheme if the realized gains do not meet the theoretical expectations. Several methods to measure the extent of genetic change have been used with varying levels of success (Rye and Gjedrem 2005). The main difficulty is that change of phenotypes or individual performance over time depends on the sum effect of all genetic and environmental changes, rather than discrete and quantifiable individual effects. It is particularly challenging to estimate the initial response to selection, i.e. when selection has been applied for only one or two generations since responses may fluctuate across generations. However, when selection has been applied for several generations, responses tend to stabilize. This chapter presents documented response to selection for a number of farmed aquaculture species that have been subject to selective breeding experiments and programs across the world, highlighting the dramatic gains that can be achieved.

3.2 Atlantic Salmon

The family based breeding program for Atlantic salmon, (*Salmo salar*) in Norway was started by AKVAFORSK in 1975 (Gjedrem 2000). For the first two cycles of

T. Gjedrem, M. Baranski, *Selective Breeding in Aquaculture: An Introduction*, Reviews: Methods and Technologies in Fish Biology and Fisheries 10,
DOI 10.1007/978-90-481-2773-3_3,

Table 3.1 Genetic gain in Atlantic salmon over five generations of selection (Thodesen et al. 1999)

Trait	Improvement in selected over wild (%)
Growth rate	+113
Feed consumption	+40
Protein retention	+9
Energy retention	+14
FCR[1]	–20

[1]Feed conversion ratio or kg feed per kg body weight produced.

selection, growth rate was the only criterion. In later generations, the breeding goal has gradually been extended and today includes age at sexual maturity, improved resistance to diseases and a number of traits related to product quality. For the first six generations of selection for growth rate, an average realized selection differential of 14% per generation was achieved (Gjerde and Korsvoll 1999), which corresponds well to other studies reporting 12–15% response to selection per generation in growth rate for this species (Gjerde 1986; Flynn et al. 1999).

A study conducted by AKVAFORSK compared the performance of Atlantic salmon selected for five generations with the performance of wild fish from the Namsen river, which constituted an important part of the base material for the selection program. A summary of the results is presented in Table 3.1. Strikingly, the growth rate of the selected material was more than twice that of the wild Namsen stock. The selected line had 40% higher feed intake, not surprising given its much faster growth. A particularly interesting finding was that both protein and energy retention was increased in the selected line, even though these traits were not selected for directly. This clearly demonstrates that selecting salmon for faster growth rate also improves efficiency in utilization of both protein and energy in the feed through a correlated selection response. The faster growth and higher protein and energy retention resulted in an overall 20% better feed conversion ratio (FCR) as compared to the Namsen stock. Since feed represents more than 60% of the production cost in salmon farming, a 20% reduction in feed requirements has had a dramatic effect on the profitability in the industry.

In addition to growth and feed conversion efficiency traits, selection for resistance to bacterial and viral diseases has become an important part of the Atlantic salmon breeding programs. A bidirectional selection experiment was carried out for resistance to infectious pancreatic necrosis virus (IPNV) using the selected Aqua Gen strain and a wild strain from the Rauma river as a control. Challenge tests were performed with fry in freshwater (Fig. 3.1) and smolts in seawater. The results showed that progeny from families selected for low resistance to IPNV suffered twice as high mortality (66.6%) than the progeny from families selected for high resistance (29.3%). The control group had an intermediate level of mortality between the selected groups (42.2%) (Storset et al. 2007). The survival of families in controlled IPNV challenge tests has been shown to have a high correlation with

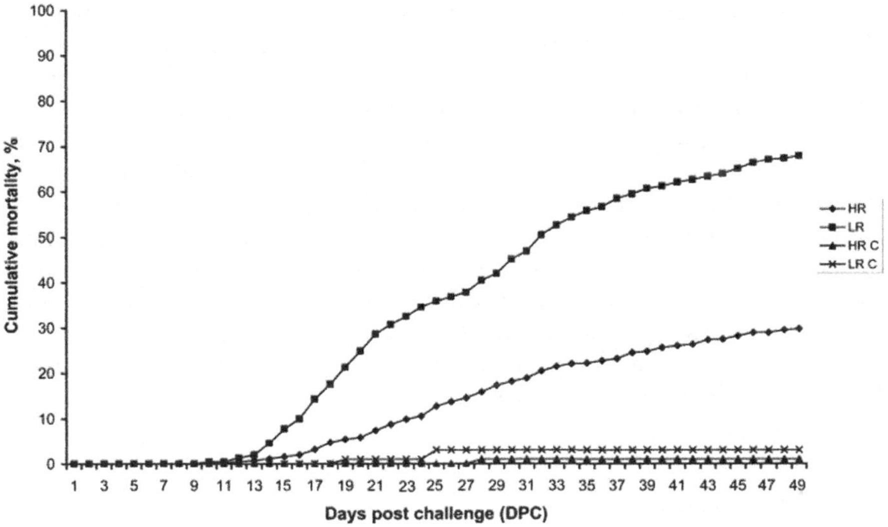

Fig. 3.1 Results from the fry stage challenge test for IPNV resistance showing average cumulative mortality in seven replicate tanks of the high resistance (HR), the low resistance (LR) and wild strain (Rauma) groups, and their respective unchallenged controls (C). Reproduced from Storset et al. (2007) by permission of Elsevier

survival of the same families during field outbreaks affecting fish at the fingerling and post-smolt stages.

An Icelandic experiment with sea ranching of Atlantic salmon demonstrated the selection response for the rate of return to the point of release (Jonasson 1994). In total, 20,720 smolts from control groups were released from five sites together with 16,286 smolts from selected parents of families with high return frequency. The return frequency was 2.2% for control groups and 2.8% for selected groups, a difference of 27% that was highly significant, Fig. 5.6.

3.3 Rainbow Trout

Rainbow trout (*Onchorhyncus mykiss*) have been farmed for more than a century and this species is now produced in more than 50 countries. Selection for growth rate has been very successful, with estimates of genetic gain ranging from 10% (Kincaid et al. 1977) to 13% per generation (Gjerde 1986).

A breeding program in Finland has selected for growth rate and early sexual maturation over two generations (Kause et al. 2005). The response to selection for faster growth rate was 7% per generation. Culling early sexually maturing males prevented an increase of early sexual maturation in one population while in another population the frequency of early maturing females increased.

A selection experiment in California for early spawn date (Siitonen and Gall 1989) demonstrated considerable genetic variation for the trait, with an estimated heritability (expressing the additive genetic proportion of the observed phenotypic variance) of approximately 0.50. The two year-classes behaved similarly and the average response in the selected stocks was seven days per generation of selection.

Like in Atlantic salmon, the IPN virus has proven problematic for rainbow trout in some areas. In Japan, resistance to IPN in rainbow trout has been improved dramatically through selection. Okamoto et al. (1993) reported an average mortality in a resistant strain of 4.3% compared with 96.1% in a highly sensitive strain.

A two-way selection experiment for muscle lipid content was performed in rainbow trout and carried out at INRA in France (Quillet et al. 2005). One year old pan-size fish were mass selected using a non destructive measurement (Disttell Fish Fatmeter) and values were corrected for absolute weight. The difference between the fat line (29.6% dry matter content) and lean line (25.6%) after two generations of selection was statistically significant.

3.4 Coho Salmon

At the University of Washington in Seattle, a selection program was carried out with coho salmon (*Oncorhynchus kisutch*) to improve traits of importance for the saltwater net-pen industry. The program produced two lines of salmon, maturing in odd and even years. After four generations of selection, growth rate increased by around 60 or 15% per generation (Hershberger et al. 1990). The traits under selection, pre-smolt and post-smolt growth and survival, continued to show positive response after nine and 10 generations of selection (Myers et al. 2001). In a later study, when selection had been performed for 16 generations, Neely et al. (2008) found that fingerlings of selected fish utilized dietary lipids for energy while saving protein for growth, while unselected fish deposited dietary lipids as body fat. It was concluded that selection over 16 generations for growth also resulted in change in feed efficiency and energy allocation.

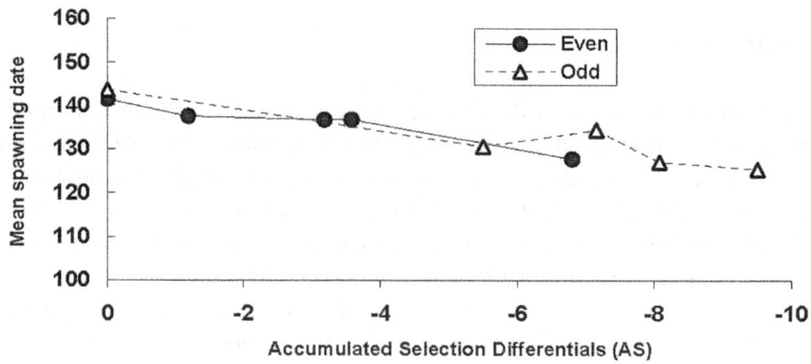

Fig. 3.2 Mean spawning date of females of two populations of coho salmon over accumulated selection differentials (AS). Reproduced from Neira et al. (2006b) by permission of Elsevier

Early spawning date has been a focus of research in Chile and Neira et al. (2006a, b) carried out a selection program in two populations of coho salmon (even and odd year class spawning). Response to selection for early spawning date is shown in Fig. 3.2. Mean spawning dates were 13 and 15 days earlier after four generations of selection in the even and odd year class, respectively. This represents an average phenotypic response to selection for onset of spawning of –3 days per generation for the two populations of coho.

3.5 Tilapia

Tilapia (*Oreochromis niloticus*) are a group of fish species native to Africa that are now farmed in more than 140 countries, and production is increasing at a rate amongst the highest of all fish species.

The three first reports of selective breeding in tilapia produced discouraging results (Teichert-Coddington 1983; Hulata et al. 1986; Huang and Liao 1990). All applied individual selection for growth rate with negligible response to selection. This shows that individual selection may be inefficient if environmental conditions are not standardised, and if the effective number of breeders (Ne) is not large enough to avoid inbreeding.

In 1988 a major selection experiment with Nile tilapia was initiated in the Philippines by ICLARM (now World Fish Center) and AKVAFORSK, in collaboration with several national research institutions in the country. The project was named GIFT (Genetic Improvement of Farmed Tilapias). The base population was formed by crossing eight strains; four wild strains from Africa and four farmed strains from the Philippines that had been under production for around 20 years. Initial strain comparison showed that the fastest growth rate was seen in the wild strains from Egypt and Kenya (Eknath et al. 1993). The strain comparison work was followed by initiation of a family-based selection scheme for growth rate, which during the five cycles of selection demonstrated an average response of 17% per generation (Fig. 3.3). The generation interval was kept at one year in the program, which resulted in the growth rate of the selected GIFT line nearly doubling over a period of only five years. Interestingly, there was no tendency of decreasing genetic gain over the years, since the highest response was obtained in the fifth generation of selection followed by the response obtained in the first generation.

Other studies in Nile tilapia have shown similar impressive responses to selection for growth rate. Bolivar (1999) reported an average selection response per cycle of 12% over 12 generations of within-family selection.

In Domasi, Malawi, Maluwa and Gjerde (2007) presented results from a selection program for harvest weight in *Oreochromis shiranus*. The realized response to selection was 6.6% per generation, a relatively low response that can be explained by low selection intensity.

Substantial genetic variation has also been documented for other traits in tilapia, including age at sexual maturation. Sexual maturation at young age/small sizes

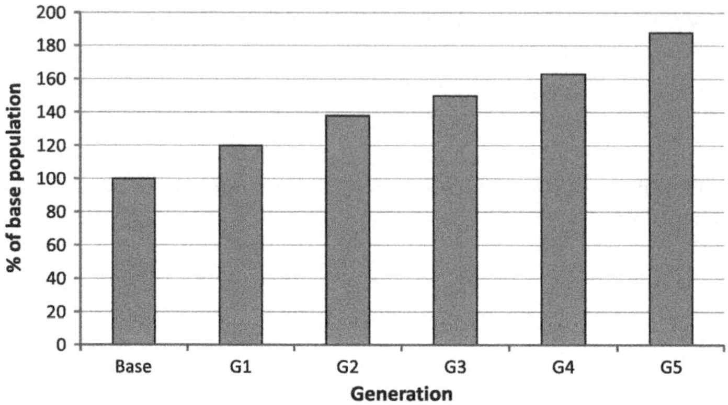

Fig. 3.3 Selection response in the GIFT project for increased body weight at harvest, measured as the percentage of the base population mean. For each generation, the response is calculated by comparing progeny of selected parents and progeny of parents with average breeding values. Reproduced from Bentsen et al. (2003) by permission of Elsevier

is a considerable problem in tilapia farming. In a selection experiment using the GIFT strain of Nile tilapia, a genetic response was obtained after one generation of selection for high (57%) and for low (34%) frequency of early sexual maturation (Longalong et al. 1999). The authors concluded that it is possible to reduce the frequency of early sexual maturation in Nile tilapia through selection.

3.6 Carp

In the 1970s, several selection experiments were carried out in Israel applying individual selection to improve growth rate in common carp (*Cyprinus carpio*). Moav and Wohlfarth (1973, 1976) summarised the results and concluded that although selecting for slow growth rate produced a response, selection for increased growth did not produce positive results. More recent studies have, however, shown that growth is a highly heritable trait (Kocour et al. 2007).

For the Indian rohu carp (*Labeo rohita*), a family-based selection experiment has been carried out at the Central Research Institute of Freshwater Aquaculture (CIFA) in India in collaboration with AKVAFORSK, with a major focus on selection for improved growth rate. The base population for the program was sampled from five Indian river systems and one farmed stock. Four generations of selection were performed with rohu reared in monoculture while for two generations rohu were reared in polyculture together with the carp species mrigal and catla. Harvest body weight differences in offspring from selected and average parents are shown in Fig. 3.4. The overall response was 29.6% per generation, which is extremely high and shows that there was substantial genetic variation for growth rate in rohu carp. The response was slightly higher in monoculture (30.6%) compared with polyculture (27.0%)

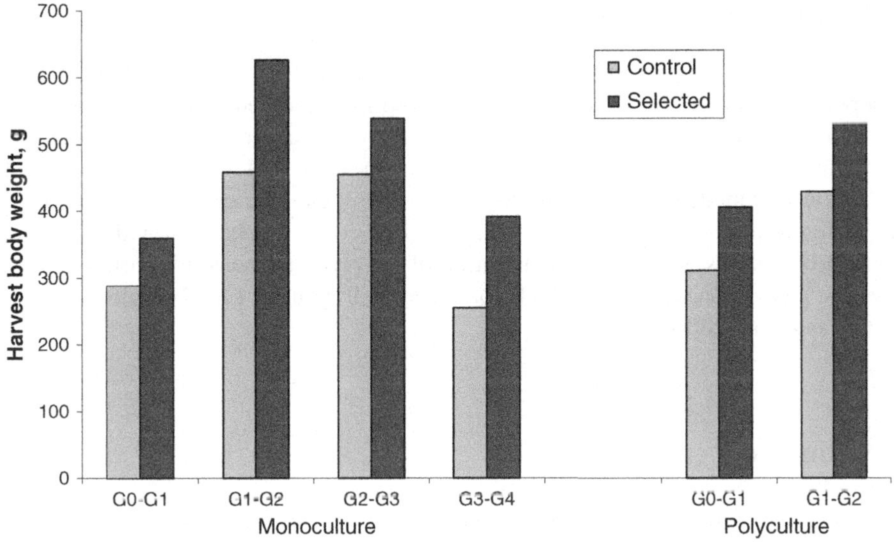

Fig. 3.4 Response to selection for growth rate in rohu carp (Mahapatra et al. 2004)

(Mahapatra, et al. 2004). These results show that it is possible to double the growth rate of rohu carp in less than four generations by selecting for only growth rate.

Silver barb (*Barbodes gonionotus* Bleeker) is another carp species that has been the focus of selection for improved growth rate. At the Bangladesh Fisheries Research Institute, a selection experiment applying individual selection was performed, with a base population consisting of a 3×3 diallel cross between three unrelated stocks. The response to selection over three generations for faster growth rate was on average 12% per generation (Hussain et al. 2002).

3.7 Channel Catfish

Farming of channel catfish (*Ictalurus punctatus*) has long traditions in the southern states of the USA, and selection for improved growth rate has been performed with excellent results. In Georgia, a bidirectional selection experiment was carried out for growth rate during the 1970s (Bondary 1983). Weight changes, measured as deviations from the control line, were around 20% in both directions in tests carried out in tanks and cages. More recently, research by Dunham (2006) reported that six generations of selection for improved growth rate has lead to an overall response of approximately 80%, corresponding to an average of 13% per generation. The authors also concluded that correlated responses to selection affected several other traits, both positively and negatively.

3.8 Sea Bream

In a breeding program for sea bream (*Sparus aurata*) in Greece run by Kego S.A. in cooperation with AFGC, selection is performed for increased growth rate, reduced incidence of deformities and improved external pigmentation. Since sea bream is a hermaphroditic species involving change of sex from males to females, the first selection was made among males only. The genetic gain for growth rate, expected to represent approximately 50% of the gain that could be obtained with selection among both sexes, was 12.1%. Estimation of the predicted selection differential in the first generation was 22%, which compared well with the actual results obtained (Thorland et al. 2006).

3.9 Shrimp

Production of shrimp worldwide has increased rapidly in recent decades. The Asian species *Penaeus monodon* previously dominated production but more recently the American species *P. vannamei* has overtaken *P. monodon*, and is now also widely produced in Asia. Since the late 1980s, viral diseases have had disastrous effects in all the main shrimp producing countries. Among these, white spot syndrome virus (WSSV) is one of the most devastating and continues to cause substantial mortality in many countries.

In Colombia, a multi-trait breeding program operated by Ceniacua in cooperation with AFGC has selected for increased growth rate, overall pond or tank survival and survival for WSSV based on controlled disease challenge tests. Two sub-populations have been selected, one for four and the other for five generations. The average selection response per generation has been 4.2% for growth rate, 5.7% for survival (under absence of specific pathogens) and 1.7% for WSSV (Gitterle et al. 2006). The most consistent response between sub-populations and generations was for growth rate. In spite of an unfavourable, negative genetic correlation between growth rate and WSSV, the family based multi-trait selection provided simultaneous genetic gain for all traits. However, although some response has been achieved for WSSV resistance, it is still too low to support an economically viable industry in heavily WSSV affected areas (Cock et al. 2008).

At the Oceanic Institute in Hawaii, selection for growth rate and resistance against Taura syndrome virus (TSV) in *P. vannamei* was carried out. Fjalestad et al. (1997) reported a response to selection of 12.4% for higher survival after TSV challenge tests and a simultaneous response in growth rate of 4.4%. In 1998, two separate breeding lines were established. One line was selected for growth rate only and a second line was selected on an index weighted 70% for TSV resistance and 30% for growth rate. After one generation of selection, the high growth line was on average 21% heavier than the unselected control line. In the TSV resistant line there was an 18% increase in survival for TSV between selected and control shrimp while the growth rate was 5% lower (Argue et al. 2002).

Divergent selection for high and low growth rate in *P. japonicus* was carried out at CSIRO, Australia. The direct response to one generation of selection averaged 11%, with 8% for high growth rate and 13% for low growth rate (Hetzel et al. 2000).

3.10 Oysters

Oysters have the highest production of all aquaculture species worldwide, and have a long history of farming in Asia, Europe and America. Most common of all the species farmed is the Pacific oyster (*Crassostrea gigas*).

In a breeding program run on the west coast of the USA, selection has been undertaken for increased live weight yield, which is a combination of individual growth rate and survival. Response to selection ranged from a 0.4 to 25.6% improvement in yields of families from selected broodstock compared with families from wild broodstock. The average response to selection over seven trials resulted in a genetic gain per generation on average 9.5% (Langdon et al. 2003).

Four breeding lines were used in a selection experiment in Australia and compared with two non-selected groups over two generations of selection. The average increase in weight for the selected lines over the average controls was 18 or 9% per generation (Nell et al. 1999). Both Newkirk and Haley (1983) and Barber et al. (1998) obtained higher selection response for growth rate, 17 and 20%, respectively.

In Sydney rock oyster populations (*Saccostrea glomerata*) in New South Wales, Australia, a parasite (*Marteilia sydney*) frequently causes high mortality in farming environments. After two generations of selection, mortality was reduced from 85.7% in the control group to 63.5% for the most improved breeding line. This represents a reduction in mortality of 22% after two generations of selection (Nell and Hand 2003).

Dramatic improvements in growth rate have also been observed in New Zealand where selection for growth rate over two generations resulted in a 20% increase in growth rate per generation. In subsequent generations, more selection pressure was placed on shell shape (Nick King pers. comm.).

3.11 Scallops

Relatively few genetic studies have been performed in scallops, a species with emerging aquaculture focus. In La Paz, Mexico, Ibarra et al. (1999) studied the response to selection in catarina scallop (*Argopecten ventricosus*) and found significant genetic gain in total weight and shell width. They concluded that growth improvements in catarina scallop can be achieved by selecting the top 10% of individuals each generation for total weight which will then result in an estimated increase in weight of 8 to 10 g per generation, or a 16% genetic gain.

3.12 Genetic Improvement in Aquatic Species Compared to Terrestrial Livestock Species

Efficient breeding programs in farm animals started back in the 1940s for poultry, cattle, pigs, and later for sheep and goats. The genetic gains obtained vary widely from species to species and for the traits in question but tend to range from about 1 to 5% per generation. Highest genetic gain has been obtained for milk yield in dairy cattle and for growth rate in broilers. This is a much lower response compared with what has been demonstrated for growth rate in aquatic species where 10–20% genetic gain is often obtained.

The main reasons for the large difference in response to selection and genetic improvements between farm animals and aquatic species are:

- Fecundity is much higher in fish and shellfish compared with terrestrial livestock species, making it possible to practise much higher selection intensity in aquatic species
- Standard deviations for body weight are two to three times higher in fish and shellfish compared to livestock species (Table 3.2).

Given that heritabilities for traits in aquatic species are generally similar to those observed in terrestrial species, the much higher responses to selection that are observed in aquatic species are a function of the higher selection intensity that is practiced in them and the larger coefficient of variation. This is discussed further in the following chapter.

3.13 Summary and Conclusion

Most estimates of genetic gain obtained for growth rate in aquatic species are between 10 and 20% per generation of selection, remarkably high figures compared to corresponding values for terrestrial livestock. In practice, this means that it is possible to double growth rate in four to seven generations. Both in Nile tilapia and

Table 3.2 Body weight averages and coefficients of variation (CV) in different farmed species

Species	Body weight	CV[1]	Reference
Atlantic salmon, kg	6.61	19	Rye and Refstie (1995)
Rainbow trout, kg,	3.41	21	Gjerde and Schaffer (1989)
Rohu carp, kg	0.30	31	Gjerde (pers. comm.)
Shrimp, *P. vannamei*, g	20.3	20	Gitterle et al. 2005
Broiler, kg	1.51	8	Rensmoen (pers. comm.)
Pigs, age at 100 kg	151	10	Sehested (pers. comm.)
Cattle, bulls, kg	440	7	Steine (pers. comm.)

[1]Coefficient of variation (CV) = (standard deviation/body weight) × 100.

Atlantic salmon growth rate has been doubled in six to seven generations of selection. Results obtained for rohu carp are even more impressive, with a doubling of growth rate in three and a half generations! Genetic gain has also been obtained for other economically important traits such as disease resistance, age at sexual maturation and date of spawning.

These genetic gains obtained for growth rate in aquatic species are in general four to five times higher than what is commonly found for terrestrial livestock species. There are two key reasons for these differences:

- Aquatic animals have a much higher fecundity in both sexes
- The phenotypic and genetic variation in economically important traits is much larger in aquatic animals compared with terrestrial livestock.

Response to selection seems to be universal across environments and species types, as similar selection responses have been observed in coldwater and tropical species, as well as in carnivorous and herbivorous species.

The economic benefits resulting from selective breeding are remarkable, and are described in more detail in Chapter 16. These economic benefits are primarily seen through the reduction in production costs due to faster turnover rate through higher growth rate, reduced maintenance requirements, increased retention of energy and protein, and better feed conversion ratio.

The results reported above have resulted from seven to eight generations of selection, and there are no indications that similar responses can not be maintained in the future (discussed further in Chapter 4). If breeding programs are well designed and the accumulation of inbreeding is controlled, there are no foreseeable plateaus in the amount of exploitable genetic variation. Hill (2008) states that 'Even if we do not understand all the results, it is clear from what has been presented that breeders have been highly effective in producing very large genetic changes over very long periods, and that there is reason to expect continued rapid change'. Such a statement is even more encouraging in the light of increased focus on environmental issues, since selective breeding not only results in economic improvements for the farmer, but also results in much better utilisation of natural resources. As productivity increases, more meat is produced per kilogram of feed, litre of water and unit of land area, meeting the increasing demands of growing populations.

Chapter 4
The Theoretical Basis for Breeding and Selection

4.1 Introduction

One of the pre-eminent professors in the field of animal breeding, A.B. Chapman, once described animal breeding sciences as follows: 'Livestock means different things to different people; seldom do we think of them as they are biologically, each a mass of millions and millions of tiny cells organised into what we call a cow or a sheep or a hog. Seldom do we pause to consider that these cells have reached this stage of organisation as a result of many cell divisions which began, as far as each particular animal is concerned, with one cell (as small as a particle of dust) formed by the union of two – the sperm from the sire and the egg from the dam.

The observable differences between individuals in a herd of pigs or cattle, a flock of sheep, or a cage of fish, are fundamentally the result the genetic code stored within these sperm and eggs, and the complex genetic and environmental interactions and processes that have occurred subsequently'.

Ultimately, the animal breeder has two main decisions to make. One is to decide which animals to select for the breeding population, and the other is to decide which animals to mate with each other. The theory of animal breeding revolves around finding and selecting the genetically superior animals that will produce the best progeny.

4.2 The Cell

Cells are the basic units of a body and can number in the order of ten billion in a single individual. The cell is surrounded by a thin permeable membrane that allows transport of molecules in and out of the cell. Within a cell, the fluid cytoplasm contains many organelles with different function, and the nucleus contains the DNA molecule that stores all genetic information. The biological function of the cell is very complex and extensive. Lysosomes act as garbage disposal systems in the cell, removing unwanted products of ingestion. Mitochondria are complex bodies that play a central role in the production of energy and cellular respiration. The numerous ribosomes function in the synthesis of amino acids that are transported to the cell

T. Gjedrem, M. Baranski, *Selective Breeding in Aquaculture: An Introduction*, Reviews: Methods and Technologies in Fish Biology and Fisheries 10,
DOI 10.1007/978-90-481-2773-3_4, © Springer Science+Business Media B.V. 2009

membrane by the Golgi bodies. Amino acids play central roles both as building blocks of proteins and as intermediates in metabolism.

After fertilisation, an individual begins to grow and develop as the zygote divides into two cells, then subsequently four, eight and so on. At each cleavage, the nucleus in the somatic cells (body cells) divides such that number of chromosomes and the genetic information is the same in all cells in the body. Quite early in development, the cells start to specialise and develop into different organs. The cleavage of somatic cells is known as mitosis and the daughter cells are always identical to the mother cells in terms of the genetic information they contain.

The cleavage of the sex cells is known as meiosis and has a fundamental difference to the process of mitosis. In the process of meiosis, eggs and sperm become haploids and contain only one set of chromosomes (1 N) compared to the cells in the mitosis process, which are diploid (2 N). Figure 4.1 illustrates the process of meiosis.

4.3 Basic Genetics

4.3.1 Introduction

The fundamental understanding of genetics as we know it today began back in the 1860s when Gregor Mendel carried out his experiments with garden peas, studying the inheritance of traits. From the results of these experiments, he postulated the existence of basic genetic units that we now call genes. Furthermore, these experiments led him to formulate two laws:

Mendels' first law: The Law of Segregation. Members of each pair of alleles separate when gametes are formed. A gamete will receive one allele or the other.

Mendels' second law: The Law of Independent Assortment. Two or more pairs of alleles segregate independently of one another during gamete formation.

At that time, the importance of these findings was not fully appreciated, but around the turn of the century, the work of Mendel was rediscovered. Our current understanding of genetics stems from these early experiments, and over 100 years of applied research and technological advancements.

4.3.2 Genes

The genetic instructions used in the development and functioning of all known living organisms are stored in DNA (deoxyribonucleic acid). Watson and Crick (1953) showed that the structure of DNA molecule was not unlike a ladder twisted into a right-handed double helix (Fig. 4.2). The uprights of this 'ladder' consist of

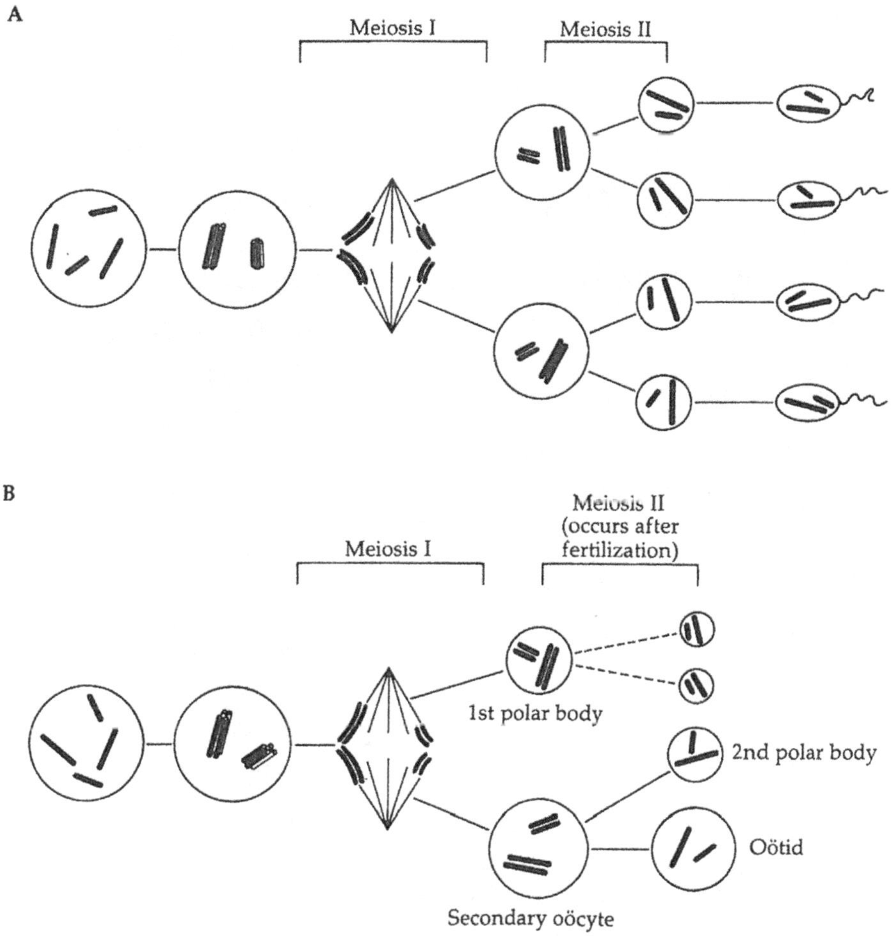

Fig. 4.1 Spermatogenesis and oogenesis in a species with two pairs of chromosomes. (**A**) In the formation of sperm, duplicated members of each pair of chromosomes come to lie side by side in four-strand configurations. Two successive nuclear divisions then result in the formation of four sperm, each with one member of each pair of chromosomes. (**B**) Meiosis in a female animal gives rise to only one functional egg from each primary oôcyte (Reproduced from L.W. Sharp 1934 with permission of McGraw-Hill, New York)

alternating phosphate (P) and deoxyribose sugar (S) groups. The cross rungs, or the steps in the ladder, consist of paired combinations of four different bases: Adenine (A), Guanine (G), Thymine (T) and Cytosine (C). Each base pair always consists of a purine (A or C) and a pyrimidine (C or T) in four different combinations: C + G, G + C, A + T and T + A. In humans, the estimated number of bases in this chain is 3.2 billion, so the DNA molecule is very long. If it could be stretched out to its fullest extent, the human DNA molecule would be around 2 metres long!

The genetic instructions that the DNA molecule contains affect all aspects of the development, metabolism, behaviour, maintenance and reproduction of an organism

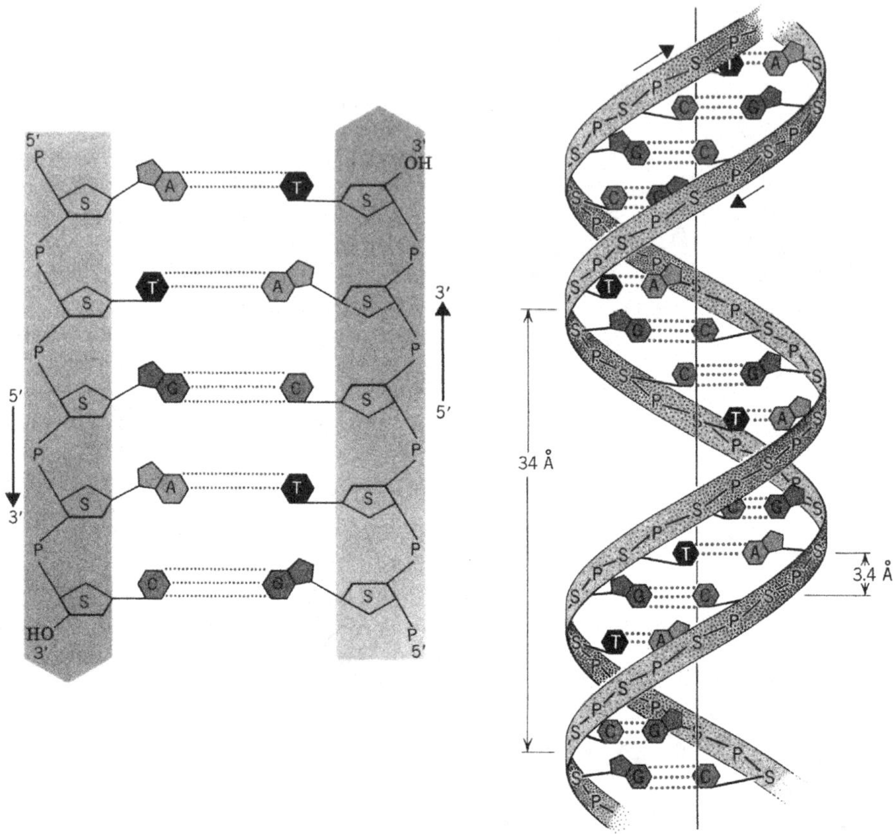

Fig. 4.2 Left: Base pairing between the two DNA strands. Right: The double helix structure of DNA. Reproduced from Gardner et al. (1991) by permission of John Wiley & Sons, Inc.

from conception to death. With all the possible sequences of base pairs along the DNA molecule, the genetic information that is stored along the molecule is almost infinite. Indeed, if the information stored along the DNA molecule in an animal cell could be documented on paper, one million pages would be needed and the stack of books would be 70 metres high!

The DNA molecule located in the cell nucleus is organised into pairs of structures called chromosomes, with one being inherited from each parent. Unless the cell is undergoing division, these chromosomes are not visible. Chromosome numbers vary widely between species, some examples of diploid chromosome number from aquaculture species are 28 in mussels, 118 in prawns, 48 in grass carp, 60 in rainbow trout, and 58 in Atlantic salmon. The number of chromosomes is considered to be stable within a species, however in several fish species, number of chromosomes has been found to vary (Grammeltvedt 1975).

Along the DNA molecule lie individual genes, the basic units of heredity in a living organism. A gene contains both coding sequences that determine the actual function of the gene, and non-coding sequences that determine when the gene is actively expressed. Like chromosome numbers, the number of genes also varies between species. The estimated number in humans is around 32,000. These genes fundamentally affect nearly all traits in an individual, including obvious physical characteristics such as colour, length and growth, as well as behavioural characteristics. The gene sequence itself defines the amino acids produced that subsequently form proteins.

A number of traits characterise the phenotype or appearance of an individual. A species can be defined as a group of organisms capable of interbreeding and producing fertile offspring. A population is the collection of inter-breeding organisms of a particular species that shares a particular characteristic of interest, most often that of living in a given geographic area. Within a species, there may be many strains or subpopulations adapted to local conditions as a result of natural selection.

4.3.3 Effect of Genes

The location of a gene on a chromosome is called a locus (plural loci). Allelic variation at individual loci, and the many possible combinations of alternative alleles, is what underlies the genetic variation in any given trait. There are two broad categories of gene effects: Additive gene effects occur when the combined effects of alleles at different loci are equal to the sum of their individual effects. In many cases, however, the effect of a given allele depends on the effect of the other allele present at the same locus. This is known as the dominance effect, and represents the interaction between alleles at the same locus. Interaction between alleles at different loci is known as epistasis. In general, dominance and epistatic effects are referred to as non-additive gene effects, implying that the effect of the alleles depends on interaction with other alleles. Only additive gene effects are fully transferred to the next generation in a strict and predictable way.

In the process of cell cleavage, occasionally problems may occur that lead to changes or mutations in genes. Cells contain inbuilt mechanisms to limit and repair this damage, however this is not always successful. Therefore, mutated genes may appear, and often such mutations are recessive and harmful. When they appear in duplicate (aa) in a homozygous form, they can often have serious effects on the fitness of an organism, and in extreme cases can be lethal.

4.4 Variation

4.4.1 Introduction

Variation is the raw material with which the breeder works. Animal breeding theory describes the exploitation of genetic variation between animals. If no genetic variation is present, then the breeder cannot achieve any gain. However, genetic variation has been shown to occur for almost all traits studied in terrestrial as well as aquatic animals. It is vital to characterise and measure the extent of genetic variation so that it can be subsequently exploited in an optimal manner to change the characteristics of the target population.

4.4.2 Single Gene Traits

Some traits are controlled by a single locus and show a simple Mendelian inheritance pattern, these are known as simple Mendelian traits. A number of human diseases follow such inheritance patterns, where a mutation in a single gene causes a disease such as sickle-cell anaemia, Tay-Sachs disease, cystic fibrosis and xeroderma pigmentosa. An example of a single gene trait in an aquatic species is albinism. In this case, the allele conferring albinism (yellow body and red eyes) is recessive, so must be present in two copies to be expressed (Fig. 4.3). If, for examples, both parents of a rainbow trout family are heterozygous 'Aa' (Normal colour allele and albino allele), then the frequency of genotypes in their progeny will be 25% 'AA' (normal colour), 50% 'Aa' (normal colour) and 25% 'aa' (albino).

A small number of genes may control other traits, with clear effects of individual alleles on the phenotype. Such cases often lead to individuals being classed into relatively few groups, and are known as qualitative traits. These qualitative traits are typically robust to environmental factors.

4.4.3 Quantitative Traits

Although simple Mendelian traits exist, traits of economic importance in aquaculture are usually controlled by tens or hundreds of individual genes. In addition, such traits may also be strongly influenced by enviornmental factors. These traits are called quantitative traits and have two basic characteristics:

- The phenotype is influenced by many genes
- The phenotype is partly influenced by environmental factors.

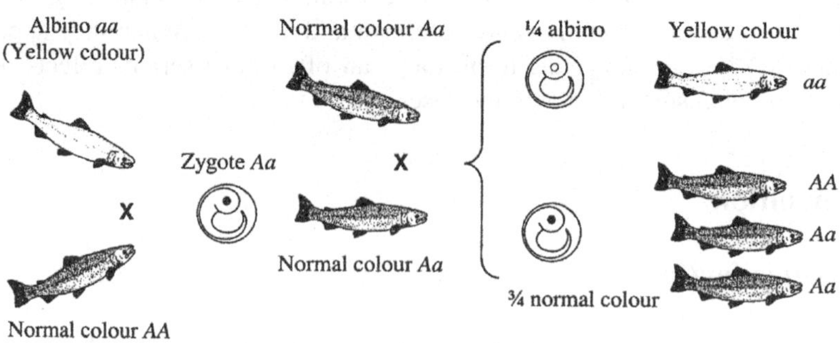

Fig. 4.3 Inheritance of albinism in rainbow trout. Reproduced from Gjedrem and Andersen (2005) by permission of Springer

A classic example of a quantitative trait that is of great interest to breeders is body size. This is a complex trait that is influenced by a series of biochemical processes that are in turn governed by many regulating genes. Different individuals may grow to the same size because of a number of different combinations of genes. In aquaculture, all traits of economic importance are quantitative and the breeding theory of fish and shellfish is based on quantitative genetics.

4.4.4 Variation in Quantitative Traits

The simplest quantitative trait is one that is influenced by two genes each with two gene variants (Aa and Bb). With four possible gametes (AB, aB, Ab and ab), these can be combined into 16 different genotypes (Fig. 4.4).

If 'A' and 'B' have the same effect (+1) and 'a' and 'b' (0), the progeny can be grouped in five classes with different genetic values. As the figure shows, most of the progeny will have the same genetic value as their parents (+2) but not the same combinations of gene variants. Only low frequency of animals will have the extreme values, 0 and 4.

According to Bentsen (2005), the approach used to construct Fig. 4.4 may be applied to any number of genes and alleles under the same simplified conditions.

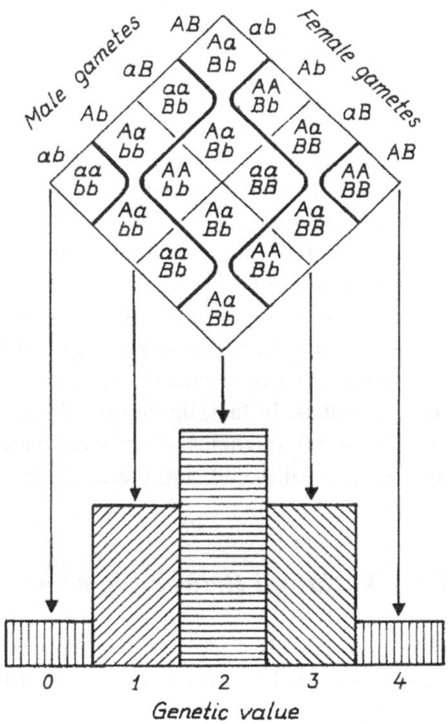

Fig. 4.4 The genotypes and distribution of genetic effects (genetic values) for a trait equally affected by two loci, each with an allele with zero genetic value (*a* and *b*) and an allele adding a genetic value of one unit to the trait (*A* and *B*) at frequencies of 0.5 in the parent generation. Reproduced from Bentsen (2005) by permission of Springer

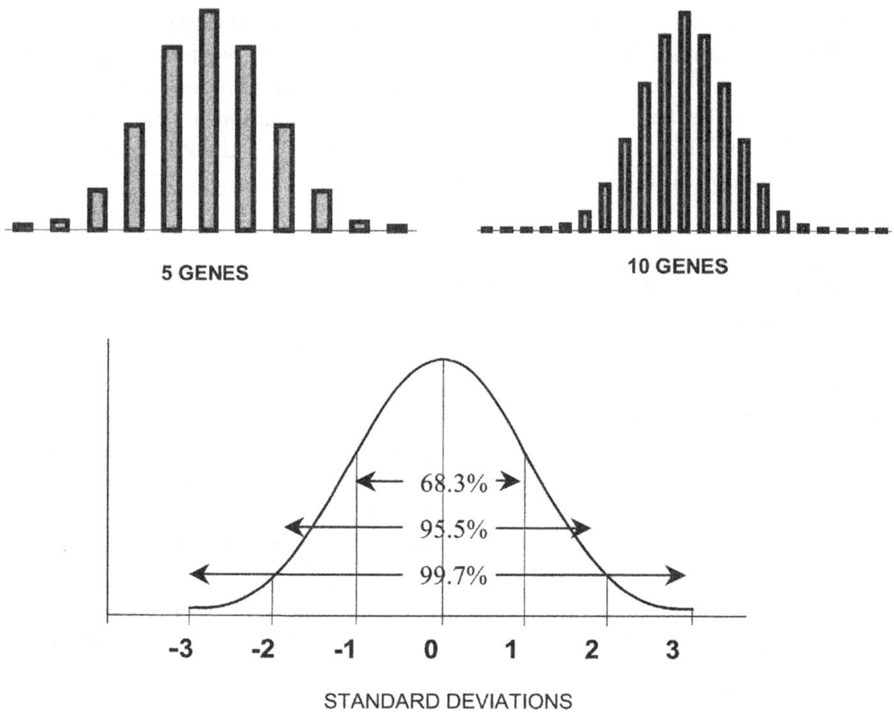

Fig. 4.5 The distribution of genetic effects in a population for a trait controlled by five or 10 genes under the same simplified assumptions as in Fig. 4.4, compared to a standard normal distribution curve. Reproduced from Bentsen (2005) by permission of Springer

This can be illustrated by the distribution of genetic effects for a trait controlled by five genes and 10 alleles, or by 10 genes and 20 alleles (Fig. 4.5). Both Figs. 4.4 and 4.5 show that as number of genes affecting a trait increases, the number of genotypes increases as well as the variation.

As the number of genes influencing a trait increases, the distribution approaches the normal distribution curve (Fig. 4.5). The normal distribution curve will cover 99.7 % of all observations within $\pm$ 3 standard deviations and even $\pm$ 1 standard deviation will cover 68.3 % of all observations. Bentsen (2005) comments that the standard normal distribution curve is open at the ends and does not include the extreme values. In fact, the normal distribution curve continues indefinitely at both ends, but the frequencies of values outside three standard deviations from the mean are rare (in total 0.3%), and increasingly rare for the extreme values.

4.4.5 Variation Between Species

There is relatively little variation between species in protein coding genes, since such genes need to be conserved in order to function. The differences between species arise mainly from how the genes are organised on the chromosomes, and how the gene sequences regulate the activity of the protein coding genes.

Variation in chromosome numbers is one of the most pronounced and obvious differences in genomic organisation between species. In addition, chromosomes can have quite different structure and content. The magnitude of these differences often implies that crosses between species do not produce any progeny. However, it has been observed that crosses between related species can produce viable progeny, but typically such progeny will not be fertile. A well known example is the mule, a cross between a horse and a donkey.

Crosses between various salmonid species have been attempted, resulting in relatively high survival among Atlantic salmon, sea trout, brown trout and Arctic charr. On the other hand, crosses between any of these species and the rainbow trout, which is from the Pacific, did not result in any surviving progeny (Refstie and Gjedrem 1975).

4.4.6 Variation Within Species

Although all individuals within a species have fundamental genetic similarity, some degree of genetic variation is always present between strains as well as between individuals. Variation between wild strains occurs due to local adaptation to different environmental conditions, and domesticated strains will vary according to the type of selection that is applied. However, evidence suggests that for most quantitative traits, the main component of variability will be found at the within – strain level.

This has been experimentally demonstrated in Atlantic salmon (Gunnes and Gjedrem 1978). Progeny from 37 wild strains of Atlantic salmon over three years classes were reared under hatchery conditions, and smolts subsequently reared in sea cages for two years. At harvest, body weight varied from 4.96 to 5.76 kg between the three different year classes. The variation between strains accounted for 7–8.6% of the total variation in body weight, hence more than 90% of the overall variation in final body weight occurred within strains. The average difference between the 19 strains tested in year class 1973 was 1.9 kg in body weight, while the difference was 2.6 kg among 13 full-sib families representing strain no. 7 (Fig. 4.6). This shows that there was significant variation between both strains and families. It is interesting to note that the differences were larger between families within a strain than the strain variation.

4.5 Estimation of Variation and Covariation

4.5.1 Mean and Standard Deviation

For traits of economic importance in all farmed aquaculture species, variation exists among species, between strains and populations, and between animals within

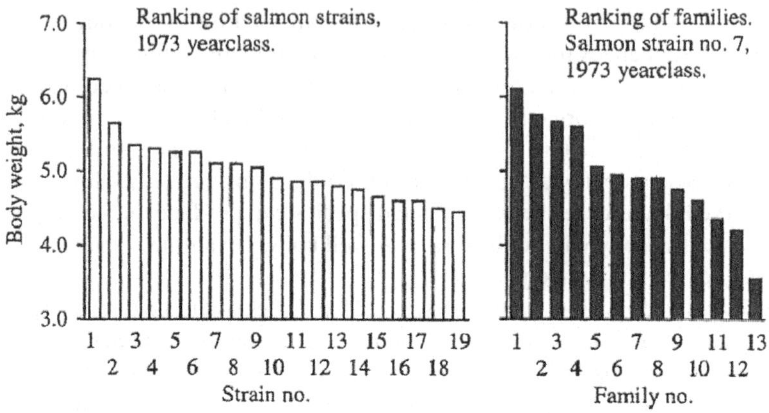

Fig. 4.6 Variation in body weight between Atlantic salmon strains and families. Reproduced from Gjedrem (1979) by permission of Tun Forlag AS.

populations. For a number of reasons it is important to be able to accurately quantify this variation. For selective breeding in aquaculture, it is of particular interest to focus on variation in key traits like body weight, body length, fat content, colour etc.

An animal's physical appearance or the measures of its traits are called the phenotype (P). This phenotype is a result of genotype (or sum of genetic effects) of the animal (G) and the sum of the effects that the environment has had on the animal over its lifetime (E). This can be expressed as follows:

$$\text{Phenotype (P)} = \text{Genotype (G)} + \text{Environment (E)} \tag{4.1}$$

In order to characterise a population, the mean or average ($\bar{x}$) describes the value of a trait in a population compared to the value in other populations. This is of particular interest when it comes to choosing breeders from different populations. To characterise a trait in a population, however, the average alone is insufficient. Some measure of the amount of variation among the animals in the population is also required. The measure used to describe this variation is called the standard deviation and is symbolised by sigma (σ).

Estimation of average ($\bar{x}$) and standard deviation (σ) is as follows:

$$\bar{X} = \text{sumX}/\text{N} \tag{4.2}$$

where X is the measurement of a trait in a number of animals (N).

The variance (σ^2) of a trait is estimated by the sum of squared deviations from the average, divided by N–1:

$$\sigma^2 = \text{sum of}\,(\text{X} - \bar{X})^2/\text{N} - 1 \tag{4.3}$$

This equation can be rewritten as:

$$\sigma^2 = [\text{sum of } X^2 - (\text{sum}X)^2/N]/N - 1 \tag{4.4}$$

The standard deviation is the square root of the variance $\sigma = \sqrt{\sigma^2}$.

If a trait is normally distributed (see Fig. 4.5), the width of the distribution will be close to six standard deviations (σ). Therefore, the average ($\bar{x}$) together with the standard deviation (σ) characterise the magnitude and variability of a trait in a population in a very descriptive manner.

An example of the use of the standard deviation: Body weight in a population is usually normally distributed. If body weight was recorded in a large population with a mean of 5 kg and a standard deviation of 1.5 kg, 68.3% of all records lie within ±one standard deviation or within (5 – 1.5) 3.5 kg and (5 + 1.5) 6.5 kg. 95.5% of all records lie within ±two standard deviations or within (5 – 3) 2 kg and (5 + 3) 8 kg, and 99.7% of all records lie within ±three standard deviations or within (5 – 4.5) 0.5 kg and (5 + 4.5) 9.0 kg.

The variance of a trait (σ^2), which is the average squared differences between each observation and the mean, is a central statistic in breeding theory.

Another useful statistical parameter is the coefficient of variation (CV), which can be estimated using the following equation:

$$\text{CV} = (\sigma/\bar{X}) \times 100 \tag{4.5}$$

Commonly, the mean and standard deviation vary in the same manner, while the CV offers a measure of variability that is independent of the mean itself. The CV is therefore useful for comparing the degree of variation in traits both between and within species.

4.5.2 Variance of a Sum

One important algebraic element in statistics is that the variance of a sum equals the variance of each of its components plus two times the covariance between the traits. Covariance indicates that two traits are correlated or vary in the same (+) or opposite (–) direction.

Earlier, it was shown that:

$$\text{P} = \text{G} + \text{E} \tag{4.6}$$

where P is phenotype of an animal, G is its genotype and E is the sum of environmental effects. Therefore, the variance of the sum of these two elements will be (Gjedrem and Olesen 2005):

$$\sigma^2_\text{P} = \sigma^2_\text{G} + \sigma^2_\text{E} + 2\text{cov}\cdot_{\text{GE}} \tag{4.7}$$

where σ^2_P is the phenotypic variance, σ^2_G is the genetic variance, σ^2_E is the environmental variance and $2cov._{GE}$ is the covariance between the genotype and the environment.

In general, the genotype and the environment are not correlated and therefore the covariance between them will be close to zero. When this is true, the equation simplifies to:

$$\sigma_P^2 = \sigma_G^2 + \sigma_E^2 \tag{4.8}$$

In order to estimate genetic parameters, it is necessary to have access to related groups of individuals like full- and half-sib families. In a breeding program, controlled mating will be implemented and a pedigree consisting of full- and half-sibs will be available. With such data, it is possible to estimate the variances in equation 4.8.

4.5.3 Genetic Variance

The total genetic value of an individual (G) can be divided into the following parts:

$$\mathrm{G} = \mathrm{A} + \mathrm{D} + \mathrm{I} \tag{4.9}$$

where A is the additive genetic value (variation due to the sum of individual gene effects), D is the dominance value (effects of interactions between alleles within loci), and I is the epistatic value (effects of interactions between loci). Assuming no interaction between these components, the total genetic variances are:

$$\sigma_G^2 = \sigma_A^2 + \sigma_D^2 + \sigma_I^2 \tag{4.10}$$

where σ^2_A is the additive genetic variance, σ^2_D is the variance due to dominance effects and σ^2_I is the variance due to epistatic effects. The sum of dominance and epistatic variance is known as non-additive genetic variance. According to Falconer and Mackay (1996), most of the interaction variance is found in the dominance component.

The genetic variance represents a valuable resource to be exploited in all aquaculture species. The additive effect is fully transmitted from parents to progeny, however this is not the case for non-additive variance. The differences between the additive and non-additive genetic variance have implications for the appropriate choice of breeding strategy, discussed further in Chapter 6.

4.5.4 Heritability

As discussed earlier, almost all traits included in breeding goals are quantitative in nature. Hence they are controlled by a large number of genes with additive effects.

The actual number of genes underlying a given trait is not known, however in most cases the number is assumed to be large with each individual gene having a relatively small effect on the phenotype. Since it is not possible to directly measure the actual effect of each individual gene, the phenotype must be used as an indirect measure with consideration of both the genotypic and environmental influences.

Heritability is an extremely important parameter in quantitative genetics. It is probably the most important statistic when developing a breeding program, as it is used to estimate breeding values and to predict response to selection. Heritability can be defined in the broad sense and in the narrow sense. Heritability (h^2_T) in the broad sense is defined as the ratio of the genetic variance (σ^2_G) to the phenotypic variance (σ^2_P):

$$h^2_T = \sigma^2_G/\sigma^2_P \tag{4.11}$$

This heritability is a measure of how much of the phenotypic variance is described by the total genetic variance. The value of the heritability in the broad sense will vary between 1 and 0. If $h^2_T = 1$, it means that all the phenotypic variance is genetic, which is not realistic for a quantitative trait, and if $h^2_T = 0$ it implies that there is no genetic variance for the trait in question, which also is rare.

Heritability in the narrow sense (h^2) describes the ratio of the additive genetic variance to the total phenotypic variance:

$$h^2 = \sigma^2_A/\sigma^2_P \tag{4.12}$$

By measuring a trait in a group of related individuals, it is possible to statistically estimate the magnitude of these heritabilities.

A heritability based on the total genetic variance (σ^2_G) will not produce correct breeding values because a part of σ^2_G could include some non-additive genetic variance ($\sigma^2_D + \sigma^2_I$), and since non-additive genetic effects are not transmitted to offspring, it should not be part of the heritability estimate when estimating breeding values. Since the total genetic variance is equal to or greater than the additive genetic variance alone, broad sense heritabilities are equal to or larger than narrow sense heritabilities ($\sigma^2_G \geq \sigma^2_A$, $h^2_T \geq h^2$).

For economically important traits in aquatic animals, heritability estimates commonly lie in the range of 0.1–0.4. Heritability is not a static parameter for traits in a population, it is specific to the population and trait in question. In Table 4.1, a selection of heritabilities are given for some species and traits. A more complete list of references is available in Gjedrem and Olesen (2005).

There are several methods for estimating heritabilities, described in detail in texts like Falconer and Mackay (1996) and Gjedrem (2005).

It is important to note that heritability estimates are strictly only relevant for the populations they are estimated from. However, if the dataset is large and containing many full- and half-sib groups, the estimated heritability is more reliable and has a more general application.

Table 4.1 Heritabilities for different traits in aquaculture species

Species and trait	Average	Heritability	Reference
Rainbow trout:			
Body weight	3.4 kg	0.21	Gjerde and Schaeffer (1989)
Fat	14.80%	0.47	Gjerde and Schaeffer (1989)
Fillet yield	63.20%	0.33	Kause et al. (2002)
Age of sexual maturity	–	0.12	Kause et al. (2003)
Survival		0.16	Rye et al. (1990)
Atlantic salmon:			
Body weight	6.6 kg	0.35	Rye and Refstie (1995)
Fat	15.60%	0.3	Rye and Gjerde (1996)
Filet yield	68.20%	0.23	Thorland pers. comm.
Flesh colour	7.7	0.47	Rye et al. (1994)
Texture	9.7	0.26	Refstie et al. (1999)
Furunculosis resistance	–	0.48	Gjedrem et al. (1991b)
Age of sexual maturity	–	0.2	Gjerde and Gjedrem (1984)
Survival		0.08	Rye et al. (1990)
Rohu carp:			
Body weight	440 g	0.23	Gjerde et al. (2004)
Survival		0.16	Gjerde et al. (2004)
Tilapia:			
Body weight	181 g	0.3	Ponzoni et al. (2003)
Survival	–	0.08	Eknath et al. (1998)
Shrimp:			
Body weight	16 g	0.17	Gitterle et al. (2005)
Survival	–	0.04	Gitterle et al. (2005)
Oysters:			
Body weight	–	0.16	Evans and Langdon (2006)
Body weight	24 g	0.28	Jarayabhand and Tavornyuikarn (1995)
Scallops:			
Body weight	–	0.46	Ibarra et al. (1999)
Total weight	61 g	0.21	Crenshaw et al. (1999)

4.5.5 Environmental Variance

In most cases, a large number of environmental factors influence a quantitative trait measured in an individual. These environmental factors may be divided into systematic and random environmental effects (Fig. 4.7).

In a breeding program, a major goal is to obtain the best possible estimate of the genotype of each animal. This estimate is termed the animal's breeding value, and is typically predicted based on phenotypic records of the animal itself and its close relatives. The influence of environmental factors tends to mask the underlying genotype, and should therefore be minimised in order to obtain the most accurate estimates of the animals' breeding values. This can be achieved in a number of ways. Firstly, the different genetic groups to be compared (e.g. stocks or families)

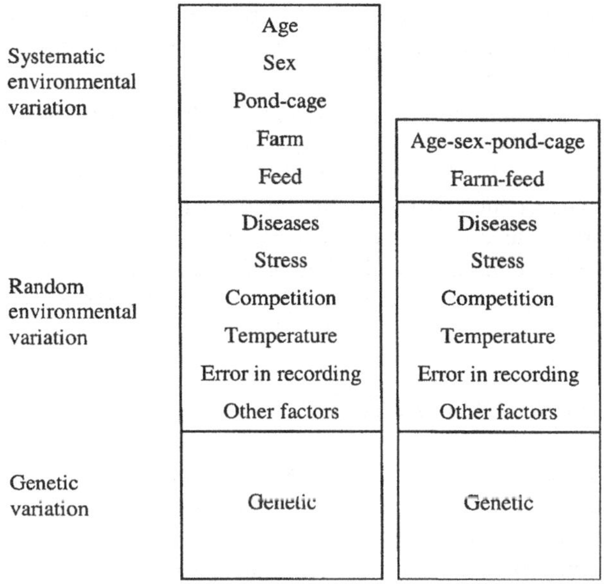

Fig. 4.7 Important sources of genetic and environmental variation in traits. The left column describes the total variation in a trait and the right column describes variation after adjustment of systematic environmental factors. Reproduced from Gjedrem and Olesen (2005) by permission of Springer

should be reared under conditions as similar as possible with respect to the type of feed, water temperature, luminous intensity and light regime. In practice, this can be effectively achieved by communal rearing of families in the same tank, pond or cage.

The effect of systematic environmental factors like age, sex, pond/cage, farm and feed can be reduced by estimating correction factors and adjusting the data. The result is illustrated in Fig. 4.7, highlighting how the environmental variation is reduced after correction. These corrections are discussed in more detail in Chapter 5.

Aquatic animals, however, pose particular problems compared to most terrestrial livestock species. In almost all cases, offspring are extremely small at hatching and cannot be individually tagged or identified at this stage. As it is vital in any breeding context to know the parents of all animals, individual families must therefore be reared in separate tanks/hapas until they are big enough to be tagged. This type of rearing introduces some environmental effects common to families, known as the tank effect. Refstie and Steine (1978) estimated the tank effect on body weight for Atlantic salmon smolts to be 5% of the total phenotypic variance. However, the tank effect was reduced to less than 1% at harvest weight during the period in sea cages. Gunnes and Gjedrem (1978) concluded from these results that it is not important to replicate the families during the fresh water period in order to rank them accurately for growth rate performance during two further years in seawater.

An alternative to separate rearing of full-sib families during the early life stages with subsequent individual tagging is to use genetic markers to identify the animals later. Use of DNA markers in this way makes it possible to rear all families together from the egg stages and throughout the whole production cycle. This eliminates the common environmental factors in the breeding population. However, when it is time for recording several traits on each individual and separate family, tagging of potential breeders is indispensable. The use of molecular markers in parentage assignment and in other application is described in more detail in Chapter 14.

4.5.6 Correlations Between Traits

Many quantitative traits are associated with each other and therefore show varying levels of covariation. The underlying reasons for covariation between traits include the possibility that they are controlled by the same genes, known as pleiotropy, or that they are affected by the same environmental factors. A good example of a very close association between two traits is body weight and body length in fish. The relative degree of association between various traits can be compared by estimating the correlation coefficient. The coefficient of correlation can vary between +1 and −1. A correlation of 0 means that the traits are completely independent of each other while values of +1 and −1 means that there is a complete positive correlation between the traits or a complete negative correlation, respectively.

The covariance between trait X and Y is:

$$\mathrm{Cov}_{XY} = \mathrm{sum}(X - \bar{X})(Y - \bar{Y})/N - 1 \tag{4.13}$$

Correlation between two traits X and Y is calculated as follows:

$$r_{XY} = \mathrm{cov}_{XY}/\sigma_X \sigma_Y \tag{4.14}$$

where the covariance (cov_{XY}) between the two traits is divided by the product of the standard deviations of the same traits.

Figure 4.8 illustrates a more complex association between two traits with phenotypes P_X and P_Y, their genotypes G_X and G_Y, and the environments E_X and E_Y. Relationships represented by directional arrows indicate the genetic and environmental components that affect the phenotype, as well as the individual gene and environmental components that make up the composite genetic and environmental effects. Bidirectional arrows represent relationships where there is no clear cause and effect. The paths from Gs to Ps (h_x and h_y) are the square roots of the heritabilies.

The correlation between the genotypic effects of the two traits (r_{GxGy}) is known as the genetic correlation. It is important to know the magnitude of genetic

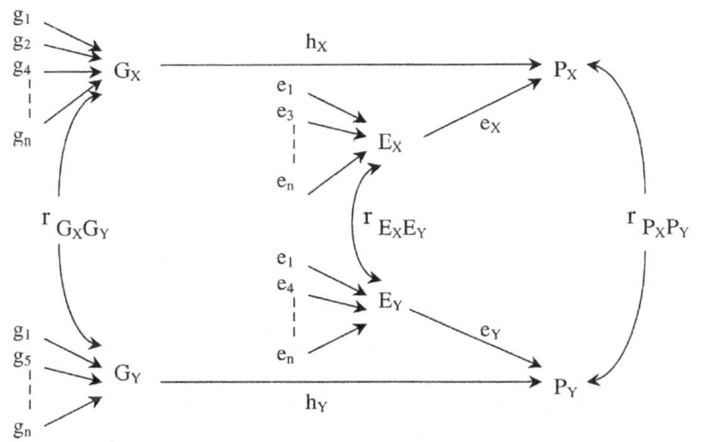

Fig. 4.8 Association between traits P_X and P_Y, which are affected by genotypes G_X and G_Y and environmental factors E_X and E_Y. g_i represents different genes and e_i different environmental factors. r_{PXPY} is the phenotypic correlation, r_{GXGY} is the genetic correlation, and r_{EXEY} is the environmental correlation. h_X and h_Y are paths from genotype to phenotype and e_X and e_Y are paths from environment to phenotype

correlations because they reveal what effect selection for one trait is likely to have on a correlated trait (see Section 4.9.6 for more details).

The association between phenotypes, or phenotypic correlation (r_P), may be due to both genetic (r_G) and environmental (r_E) correlated effects. The phenotypic correlation can easily be estimated from recorded data, however, the estimation of genetic and environmental correlations requires a large dataset with information from many relatives that have been reared under similar environmental conditions.

An example of a genetic correlation is shown in Fig. 4.9. Different full-sib families that were challenged with two bacterial diseases, furunculosis and BKD are listed along the x-axis. The families were ranked for survival after being infected with furunculosis by means of a cohabitant challenge. The similar ranking of families indicates a very high genetic correlation between resistance to these two bacterial diseases. The estimated genetic correlation was as high as $r_G = 0.81$, meaning that families with high survival after an attack of furunculosis will tend to also have high survival when infected by BKD.

Gunnes and Gjedrem (1978) studied the association between body weight and body length in rainbow trout and found the following correlations: $r_P = 0.88$; $r_E = 0.86$ and $r_G = 0.98$. All three correlations were very high, in particular the genetic correlation, indicating that the association is nearly complete. The correlation is positive, which means that as the body weight increases so does body length.

In a survey for genetic correlations between economic important traits in the literature, a range of estimates were found (Table 4.2).

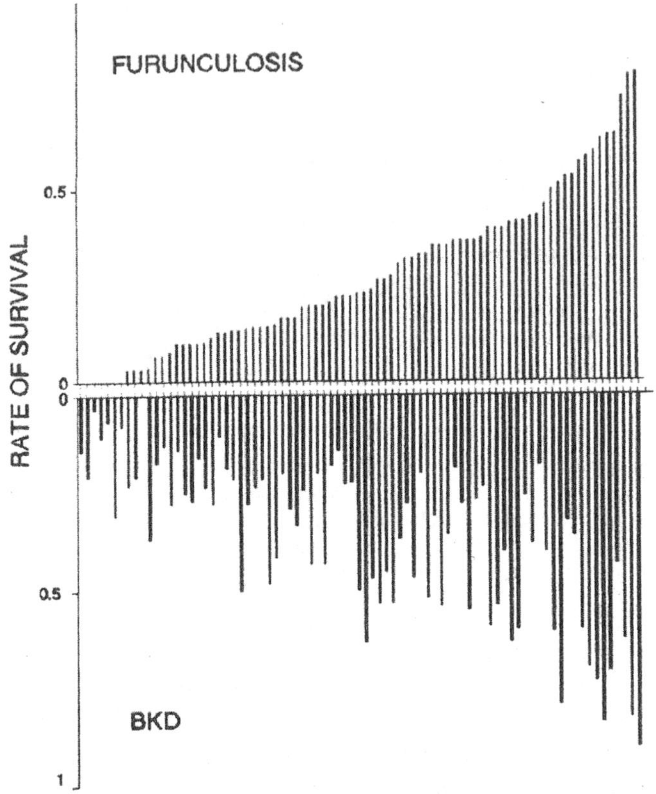

Fig. 4.9 Rate of survival for full-sib families of Atlantic salmon challenged for furunculosis and BKD. Different full-sib families are represented on the x-axis. The families are ranked according to level of survival for furunculosis. Reproduced from Gjedrem and Gjøen (1995) by permission of Wiley-Blackwell

Table 4.2 Average genetic correlations between economically important traits

Traits	No. of estimates	Genetic correlation
Body weight – survival [1]	8	+ 0.25
Body weight – fat %[2]	6	+ 0.23
Body weight – flesh colour[2]	5	+ 0.25
Fat % and flesh colour[2]	5	+ 0.10

[1]Estmates from fish and shrimp; [2]Estimates from salmonids

4.5.7 Regression

Frequently, farmers want to predict the value of one trait from the records of another trait. For example, a farmer may want to predict harvest weight from the weight of fingerlings, or predict how much feed is required to reach harvest weight.

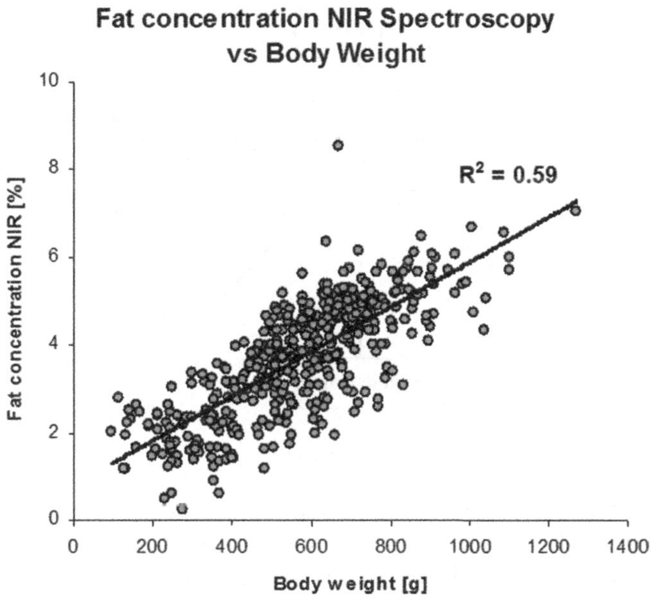

Fig. 4.10 Relationship between body weight and fat percentage in Atlantic salmon measured by NIR spectroscopy (Folkestad pers. comm.)

The appropriate statistic to make such a prediction in this case is the coefficient of regression of Y on X (b_{XY}), which can be estimated as follows:

$$b_{XY} = cov_{XY}/\sigma_X^2 \tag{4.15}$$

A simple regression equation is:

$$Y = a + b_{XY}X \tag{4.16}$$

This equation describes a straight line with the constant 'a' being the value of Y when X = 0, and a slope of 'b_{XY}'. The line passes through the averages of both traits. Figure 4.10 is an example of a regression line constructed from the observations taken of fat percentage and body weight in Atlantic salmon.

4.6 Inbreeding and Relatedness

4.6.1 Genetic Relationship

Colloquially, the relationship between individuals has been expressed by the amount of blood they share. Such statements are of course not true, but can indirectly be interpreted as referring to the fraction of genetic material that comes from shared ancestors. Relatives resemble each other because each offspring inherits half of its

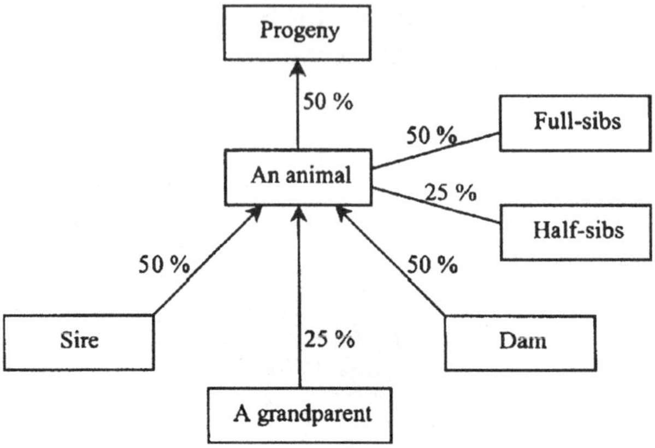

Fig. 4.11 Relationship (R) between close relatives

genetic material from each parent. The relationship between two individuals is simply the probability that they will share more genetic material than unrelated members of the same population will, because they are related by descent.

An animal can be said to be 100% related to itself. The relationship between a parent and offspring is 50% since each offspring has received half of its genetic material from each parent. In an inbred population, the relationship between parent and offspring will be higher than 50%.

The most probable relationship between full-sibs is 50% because one quarter of their genetic material will be received from the sire they have in common and another one quarter will be received from their common dam. There may be variation in relationship between full-sibs because the laws of chance may cause some pairs of sibs to receive more than one quarter of their genetic material as duplicated material from the sire or dam. On average, half-sibs are 25% related because one quarter of their genetic material is duplicated material that both animals received from the common parent (Fig. 4.11).

Wright (1921) is credited with the concept of tracing paths to establish the relationships among relatives and introduced the coefficient of relationship (R). Examples of relationships are given in Fig. 4.11. In addition, the figure could be extended to show that the coefficient of relationship between an animal and its great grandparent is 0.125 and between cousins is 0.0625.

Since the halving nature of Mendelian inheritance decreases the level of relationship per generation because of segregation, individual relationships more remote than between first cousins have little practical importance. If one of the ancestors is inbred, the relationship will be stronger.

4.6.2 Inbreeding

The broad definition of inbreeding is the mating together of individuals that are related to each other by ancestry. Such mating tends to make the offspring more homozygous than their parents were. In reality, the use of the term inbreeding is somewhat determined by the closeness of the mating and the context of the particular breeding situation. It is generally accepted that mating between parents and offspring, between full-sibs and also between half-sibs represents inbreeding, while mating of first or second cousins may not always be described as such. Plant breeders may often only consider self-fertilisation as inbreeding. In most breeding programs even weaker relationships are considered inbreeding because the results of this level of inbreeding will accumulate if mating within a closed population continues for many generations. In fish and shellfish species with very high fecundity, the risk of rapidly building up inbreeding is very high, because only a few parents are needed to reproduce each generation.

Inbreeding is commonly measured by the coefficient of inbreeding (F), a term introduced by Wright in 1921. The inbreeding coefficient is defined as the correlation between genetic values of gametes. Malecot (1948) additionally defined the coefficient of inbreeding as the probability that two genes at the same locus are identical by descent. When considering all loci in an individual, the coefficient of inbreeding represents a proportion of loci equal to F that are expected to carry genes that are identical by descent. The coefficient of inbreeding refers to an individual and expresses the degree of relationship between the individual's parents.

A population is considered to be in equilibrium (panmictic) when N reproducing individuals are mating at random with no inbreeding. It is further assumed that there are two alleles at each locus and that the total number of alleles in the population is 2 N. The probability that two gametes drawn at random carry identical genes is therefore $\frac{1}{2}$N. According to (Falconer and Mackay 1966) 'Any gamete has a ($\frac{1}{2}$N)th chance of uniting with another of the same sort, so $\frac{1}{2}$N is the probability that uniting gametes carry identical genes, and thus the coefficient of inbreeding of the progeny'.

The base population may be considered as generation 0 with no inbreeding. Progeny from this population, or generation one, will have the following coefficient of inbreeding:

$$F_1 = 1/2N \tag{4.17}$$

In generation two, the homozygote may be formed in two ways:

- A combination of identical alleles in generation one with the frequency of $F_1 = 1/2\,N$. The remaining proportion $(1 - 1/2\,N)$ has alleles that are independent in formation of gametes in generation one
- In the fraction $(1 - 1/2\,N)$, some alleles may be identical to earlier generations. This fraction is $(1 - 1/2\,N)F_1$.

The coefficient of inbreeding in generation two will therefore be:

$$F_2 = 1/2N + (1 - 1/2N)F_1 \tag{4.18}$$

and the coefficient of inbreeding in generation t will be:

$$F_t = 1/2\,N + (1 - 1/2\,N)F_{t-1} \quad (4.19)$$

The coefficient of inbreeding consists of two parts: The increase in identical alleles 1/2 N from new inbreeding and the increase due to inbreeding in earlier generations. The newly occurring inbreeding may be called:

$$\Delta F = 1/2\,N \quad (4.20)$$

The coefficient of inbreeding in generation t may then be expressed as:

$$F_t = \Delta F + (1-\Delta F)F_{t-1} \quad (4.21)$$

and

$$\Delta F = (F_t - F_{t-1})/(1 - F_{t-1}) \quad (4.22)$$

In some fish species, homozygous individuals can be produced through gynogenesis or androgenesis. Another way to rapidly increase homozygosity is through self-fertilisation, which is commonly used in plants and can occur in shellfish like scallops. However, even mating full-sibs and even half-sibs results in accumulated rapid accumulation of inbreeding (Table 4.3).

Table 4.3 Expected increase in the coefficient of inbreeding (F) from the systematic mating of relatives

Relationship			
Generation	Selfing$^{(1)}$	Full-sibs$^{(2)}$	Half-sibs$^{(3)}$
0	0	0	0
1	0.5	0.25	0.125
2	0.75	0.375	0.219
3	0.875	0.5	0.305
4	0.938	0.594	0.381
5	0.969	0.672	0.449
6	0.984	0.734	0.509
7	0.992	0.785	0.563
8	0.996	0.826	0.611
9	0.998	0.859	0.654
10	0.999	0.886	0.691
20		0.986	0.903

$^{(1)}$ ½ $(1+F_{t-1})$; $^{(2)}$ ¼ $(1+2F_{t-1}+F_{t-2})$; $^{(3)}$ 1/8 $(1+6F_{t-1}+F_{t-2})$

Reproduced from Falconer and Mackay (1996) by permission of Pearson Education Ltd.

4.6.3 Effective Population Size

In a closed population, the number of mating animals will be restricted. This number is known as the effective number of breeding animals or the effective population size (Ne). Ne is the number of breeding individuals in an idealised population that would show the same amount of dispersion of allele frequencies under random genetic drift or the same amount of inbreeding as the population under consideration. In a large random breeding population the rate of inbreeding would be $\Delta F = 1/2\,N$. The effective population size is related to ΔF in the same way and would therefore be obtained from the calculated ΔF as $Ne = 1/2\Delta F$. Knowing the effective population size, Ne, for any breeding structure, enables the rate of inbreeding to be calculated as:

$$\Delta F = 1/2Ne \tag{4.23}$$

In a breeding population, the two sexes are often unequally represented. The two sexes contribute equal amounts of genetic material to the following generation and therefore the number of males and number of females must be recorded. The effective number of each sex is twice the harmonic mean of the numbers, since the sampling variance is proportional to the reciprocal of the number. The harmonic mean is according to Falconer (1960):

$$1/\left[1/2(1/N_m + 1/N_f)\right] \tag{4.24}$$

where N_m and N_f are the numbers of males and females, respectively. The effective population size is according to Falconer and Mackay (1996):

$$1/Ne = 1/4\,N_m + 1/4\,N_f \tag{4.25}$$

$$Ne = 4\,N_m N_f/(N_m + N_f) \tag{4.26}$$

The rate of inbreeding is therefore:

$$\Delta F = 1/8\,N_m + 1/8\,N_f \tag{4.27}$$

Examples of calculating Ne and ΔF:

Example 1; $N_m = 15$ and $N_f = 100$ then $Ne = 52$ and $\Delta F = 0.010$
Example 2; $N_m = 10$ and $N_f = 100$ then $Ne = 36$ and $\Delta F = 0.014$
Example 3; $N_m = 5$ and $N_f = 100$ then $Ne = 19$ and $\Delta F = 0.026$
Example 4; $N_m = 5$ and $N_f = 50$ then $Ne = 18$ and $\Delta F = 0.028$

These results show that the effective population number is strongly affected when the number of each sex is very different.

4.6.4 Effect of Inbreeding on Genetic Variance

Inbreeding influences the genetic variance of a trait. According to Falconer and Mackay (1996), the variance of a trait after inbreeding will be reduced as the inbreeding increases:

$$V_F = V_G(1 - F) \tag{4.28}$$

where V_F is variance after inbreeding, V_G is the genetic variation in the base population and F is coefficient of inbreeding.

Inbreeding within a breeding population will therefore have very serious effects due to the loss of genetic variance, and will reduce both heritability as well as genetic gain in a breeding program. When homozygosity increases, productivity and fitness decreases. In species with high fecundity, particular care must be taken as only a few broodstock are necessary for reproduction in each generation.

4.6.5 Inbreeding Depression

As inbreeding increases in a population and genetic variance decreases, the animals will become more similar phenotypically. When forming a new breed, inbreeding is commonly used to standardise the breed and make the animals within each breed more uniform. However, in a breeding population, the reduction of genetic variance is serious because it will reduce the response to selection.

Another detrimental effect of inbreeding is that the increase in homozygosity tends to 'uncover' undesirable recessive genes, which may be lethal or semi-lethal. If an undesirable gene variant is occurring in a population at a very low frequency, the probability of its expression is small. Mating of relatives that have an ancestor carrying such a gene variant will dramatically increase the frequency of expression.

When planning a selection scheme, it is of interest to know the effect of inbreeding on economically important traits. Plant and animal breeders have found the effect of inbreeding to be generally unfavourable. This general reduction in performance due to inbreeding is known as inbreeding depression. The negative effect is most apparent in fitness traits like survival and reproduction, and for growth rate. Inbreeding depression is usually expressed per 10% of the coefficient of inbreeding (F). In general, inbreeding depression seems to be linearly related to the coefficient of inbreeding, meaning that rearing highly inbred animals is likely to cause large problems.

In Table 4.4, some estimates of inbreeding depression are presented for rainbow trout with varying levels of inbreeding from $F = 0.25$ to $F = 0.50$. The inbred lines were produced by successive generations of full-sib mating. The results are variable but the inbreeding depression was around 3–6% per $F = 0.10$. However, for highly inbred parents the results are highly variable.

Table 4.4 Effect of three levels of inbreeding in rainbow trout on survival and body weight

	Survival				Body weight			
	Eyed eggs,%		Fry,%		Fingerlings, g		At harvest, kg	
Inbreeding Coefficient[1]	Control	ID[2]	Control	ID[2]	Control	D[3]	Control	ID[3]
F = 0.25	95.1	8.9	81.4	9.1	12.0	9.8	2.50	11.3
F = 0.375	94.6	15.3	72.7	5.6	47.8	22.5	3.15	19.9
F = 0.50	96.8	5.8	72.6	18.6	12.2	−4.0	2.96	30.5

[1]Obtained by one (F=0.25), two (F=0.375) and three (F=0.50) successive generations of full-sib mating.
[2]Inbreeding depression (ID = Control – Inbred)
[3]Inbreeding depression [ID = (Control – Inbred) 100/control]
Reproduced from Gjerde et al. (1983) by permission of Elsevier

Fjalestad (2005) reviewed the literature concerning inbreeding depression in fish and shellfish. Of 28 references, 24 show inbreeding depression between 0.9 and 9.3% per 10% of inbreeding while the remaining four had higher values.

In a study in Coho salmon, selection was performed for growth rate and survival over nine and 10 generation for even- and odd-year breeders, respectively. Based on the population sizes, the inbreeding coefficients were estimated to be 22% for even and 23% for odd years fish (Myers et al. 2001). This study found a lack of deleterious responses from selection and argued that because of the high fertility of salmonids, there is considerable opportunity for deleterious alleles to be culled from the population.

Bentsen and Olesen (2002) discussed the build up of inbreeding in a simulation study where selection was performed for 15 generations and concluded that in general, rapid inbreeding accumulation through full-sib mating schemes often seems to result in a 5–10% depression of performance per 10% increase in inbreeding coefficient. For the design modelled with the highest rate of inbreeding, an inbreeding depression of 50–75% was possible after 15 generations. At low rates of inbreeding (1–2% per generation), inbreeding depression appeared to drop to below 5% per 10% increase of the inbreeding coefficient. This is in agreement of the conclusion of Myers et al. (2001).

Crossing inbred animals with unrelated individuals can increase heterozygosity, increase genetic variation and improve fitness. Bentsen and Olesen (2002) studied means of avoiding inbreeding through applying mass selection. They conclude that in order to maintain a low rate of inbreeding (about 1% per generation), a minimum of 50 pairs of breeders should be selected and the number of progeny tested should be restricted and standardised to not less than 30–50 progeny per pair.

Example of inbreeding depression: By mating full-sibs, the progeny will have a coefficient of inbreeding of $F = 0.25$. If inbreeding depression is 5% per $F = 0.10$ for growth rate, reduction in body weight in the inbred progeny will be $5\% \times 2.5 = 12.5\%$ compared with outbred animals.

4.7 Crossbreeding

4.7.1 Introduction

Crossbreeding may be performed by mating different strains, inbred lines and even crossing different species. The purpose of crossbreeding is to produce hybrid individuals with superior performance. Crossbreeding exploits non-additive genetic variance in traits in order to obtain hybrid vigour or heterosis. In some cases, considerable heterosis may occur even in the first generation that would otherwise require several generations of conventional selection to reach the same level.

4.7.2 Heterosis

Heterosis or hybrid vigour occurs when the performance of offspring surpass the average of its parents or the best parental strain for one or more traits. Heterosis is effectively the opposite of inbreeding depression. While inbreeding leads to homozygosity and reduced fitness, crossing inbred lines or unrelated populations leads to heterozygosity, increased fitness and higher productivity. Non-additive genetic effects are primarily responsible for heterosis, with the dominance component playing a large role. The high degree of heterozygosity that usually results from crossbreeding tends to produce a high frequency of dominance effects.

The second component of the non-additive genetic effect is epistasis which is defined as interaction between genes at different loci. It is somewhat difficult to get good estimates of epistasis, and evidence suggests that epistatic effects are generally marginal.

The most pronounced effects have been obtained by crossing strains of common carp in Israel (Moav et al. 1975) and Hungary (Bakos 1979), findings leading to the establishment of crossbreeding programs in these countries. In Israel, a large amount of data exists on the gains achieved (Wohlfarth 1993). The results of 73 comparisons between crossbreds and their parental lines indicate that 47 of the crossbreds showed a higher corrected weight gain than the superior parent, and in 15, the difference between the crossbred and the superior parent was significant. Overall, the empirical results indicate that crossbreds possessed faster growth rates than purebreds. However, this result was not universal, as Dor-70 (a line selected for fast growth rate over several generations) showed superior growth to the crossbreds. The advantage of crossbreds over purebreds is most obvious when the purebred lines are not exceptionally superior for the trait of interest. Significant heterosis has also been observed in crossing strains of rainbow trout (Gall 1975; Ayles and Baker 1983), and channel catfish (Dunham 1987).

Some examples of heterosis effects from a diallel cross between eight strains of Nile tilapia are presented in Table 4.5. The estimates of heterosis vary widely from 5.2 to 14%. The effects are not systematic, suggesting that it is not possible to predict the magnitude of heterosis effects. The E2 strain had the highest heterosis effects overall. In a dialell cross between four strains of *Oreochromis shrinanus*

Table 4.5 Mean percent heterosis for body weight at harvest across test environments in the 56 strain crosses of Nile tilapia

Sire strain[(1)]	Dam strain[(1)] E2	Gh	Ke	Se	Is	Si	Th	Tw
E2	–	3.8	9.9	11.6	9.3	3.1	9.5	5.7
Gh	4	–	−1.5	10.4	2.6	3.9	14	3.7
Ke	10.6	−1.5	–	−0.8	4.8	1.9	5.6	1.6
Se	12.3	10.3	−0.8	–	6.3	0.3	2.8	0
Is	10.4	2.8	4.9	6.7	–	0.6	7	0.4
Si	2.8	3.5	1.7	0.2	0.4	–	5.8	2.5
Th	10	13.7	5.4	2.7	6.5	6	–	−5.2
Tw	6	3.8	1.7	0.1	0.3	2.8	5.8	–

[(1)]Strains: Four wild strains from Africa collected in 1988–1989, Egypt (E2), Ghana (Gh), Kenya (Ke), Senegal (Se) and four farmed strains introduced to the Philippines in 1979–1984 from Israel (Is), Singapore (Si), Thailand (Th) and Taiwan (Tw). Reproduced from Bentsen et al. (1998) by permission of Elsevier

in Malawi, the total heterosis effect accounted for 15.3% of the total variance for harvest body weight (Maluwa and Gjerde 2006).

Although it is sometimes responsible for large effects, heterosis is not a universal phenomenon, particularly when crosses are made between species. Crosses of Atlantic salmon, brown trout, sea trout and Arctic charr produced offspring that neither survived better nor grew faster than Atlantic salmon (Refstie 1983a). Chevassus (1979) concluded from crossing species of salmonids that in most cases, the hybrids showed intermediate or at best equal performance to the best pure species. For sea bream crosses, Knibb et al. (1997) also found low heterosis.

At AKVAFORSK, a diallel crossbreeding experiment was carried out by crossing five wild strains of Atlantic salmon (Gjerde and Refstie 1984). The heterosis effect was generally low and not significant. There was positive heterosis for harvest weight in five crosses and for survival in six of the 10 crosses when they were compared with the average of the purebred strains. When crossbreds were compared to the best performing purebred strain, two crosses had superior harvest weight and survival. The highest heterosis effect was obtained for both harvest body weight and survival when the parental strains originating from the greatest distance of separation were crossed. Low heterosis in this species was also found by Friars et al. (1979). Rye and Mao (1998) studied the degree of non-additive genetic effect for growth rate in four sub-populations of Atlantic salmon. The results indicated that body weight at harvest was strongly affected by non-additive genetic effects with estimates of dominance and additive by additive epistatic effects accounting for 2–9% and 13–17% of the total variance, respectively.

In India, a crossbreeding experiment was carried out with two 3×3 diallel crosses of rohu (*Labeo rohita*). For harvest weight, the total heterosis for each of the six stock crosses was low or negative. For survival, the total heterosis was negligible and not significantly different from zero (Gjerde et al. 2002).

4.8 Purebreeding

Mating of unrelated individuals within a population or a strain is known as purebreeding. This method is commonly used in breeding programs for terrestrial livestock species as well as for fish and shellfish. Purebreeding provides good possibilities for progress when efficient selection is applied and is simple to apply in practice. It is a particularly effective strategy when the performance of one strain is good as or better than alternative strains. However, if a purebred strain is inferior for one or more traits, it is possible to introduce genes from another strain with favourable alleles.

The downside of the purebreeding approach in a closed population is the accumulation of inbreeding. In practise, it is not possible to avoid the accumulation of inbreeding in a closed population, however it is possible to keep it at a low and acceptable level. The key requirements for keeping the levels of inbreeding low in a closed population is:

- Avoid mating of close relatives (parents and offspring, full- and half-sibs)
- To keep the effective number of breeders (Ne) high, at least 50 pairs per generation.

4.9 Selection

4.9.1 Introduction

In simple terms, selection describes situation where individuals with advantageous or desirable traits contribute more offspring to the succeeding generation than others do. In the context of selection, the number of breeding animals is more important than the number of progeny produced by each animal, since animals that do not reproduce have no effect on future populations. Selection does not produce new gene alleles, but it changes the frequencies of alleles of genes with additive effects, resulting in an increase in frequency of favourable gene alleles and a decrease in frequency of gene alleles with negative effect.

Selection would be most efficient if one could measure the actual genotype of the animals, rather than the phenotype (which is the sum of genotype and environment). However, for the majority of traits it is only possible to measure the phenotype. For some traits with high heritability, measures of the phenotype will closely reflect the underlying genotype, while for traits with low heritability, phenotypic measurements reveal little about the genotype of the animal for that specific trait. Emerging molecular methods facilitating selection on basis of genotypic information (e.g. marker-assisted selection (MAS) and genomic selection) may be commonplace in the future, and this is discussed in more detail in Chapter 14.

When the aim of selection is to modify a trait in the population, the first step is to obtain records and measurements on all available animals in the population.

Selection can then be performed by choosing the animals with the best performance to be parents for the next generation.

4.9.2 Natural Selection

Individuals that are well adapted to particular environmental conditions and produce many surviving progeny show high fitness, while animals producing few progeny or progeny with low survival show low fitness. The consequence of these processes is that the population overall will become more and more adapted to the environmental conditions, because animals with high fitness will be more reproductively effective than those with low fitness. This process is known as natural selection and occurs in nature in all wild animal and plant populations. Natural selection is, however, a slow process, since environmental changes that animals must adapt to tend to occur very slowly. In addition, natural selection occurs only on an individual level, and is not influenced by the performance of relatives.

An example of how natural selection occurs is the creation of Atlantic salmon strains in Norway. After the last ice-age around 10,000 years ago, Atlantic salmon colonised the rivers along the Norwegian coast. Today, there are a number of strains of Atlantic salmon that are locally adapted to the conditions in these river systems. The strain differences are a result of natural selection caused by different environmental conditions in the rivers.

If the environmental conditions change rapidly, natural selection may not be sufficient to adapt populations to the new conditions. An example is the rapid acidification of waters that occurred in Southern Norway. Losses of fish started as early as the 1920s and the most rapid losses occurred during the 1960–1970s (Rosseland et al. 1986). Acid rainfall lowered the pH to a level below the tolerance level (pH <5) for salmonids, resulting in mass mortalities of fish in lakes, rivers and creeks in large regions of the country. However, an extensive investigation showed that there was considerable genetic variation for tolerance to acidic water in brown trout, with heritabilities (h^2) ranging from 0.09 to 0.33 (Gjedrem 1976; Edwards and Gjedrem 1979). Figure 4.12 show the large variation in survival between brown trout strains in acidic water during egg and alevin stages. These findings indicate that natural selection could allow brown trout to adapt to low water pH, however the mass mortalities observed in the wild populations suggest that acidification must have taken place too fast for the natural selection to act.

Natural selection is an important process not only in wild populations, but also when animals are domesticated. Wild animals brought into captivity will not necessarily thrive in these new conditions. Those that are best adapted to the captive environment, and show desirable phenotypes, will have a greater opportunity to reproduce and pass their genes on to the next generation.

4.9.3 Artificial Selection

In most cases, farmers will have a desire to improve their population, and therefore practice artificial selection, also known as directional selection. When additive

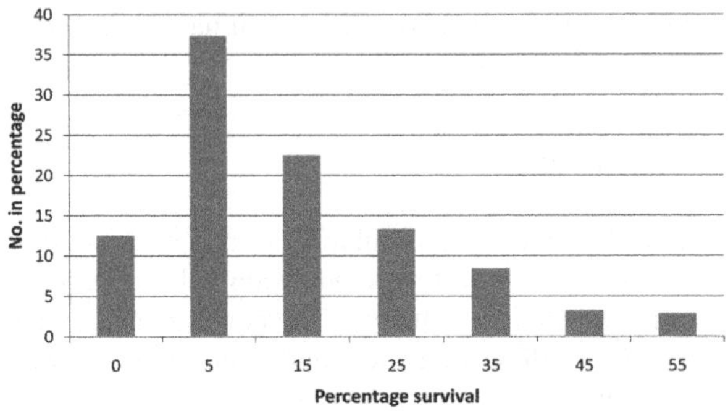

Fig. 4.12 Survival rate of different strains of brown trout in acidic water, average survival in pH= 4.7 and 5.2. Reproduced from Gjedrem (1976) by permission of The Research Council of Norway

genetic variation is present for a trait, selection is an efficient tool to improve it in a desired direction. The genetic effect of directional selection is reflected in changes in allele frequencies at loci affecting the trait.

In Chapter 3, several examples were given showing response to directional selection in several species. Some of the responses documented are very strong, implying that artificial selection is a very powerful tool to make changes in populations of aquatic species. However, this is only the case when selection is well planned and carried our carefully.

4.9.4 Predicting Selection Change

Under certain conditions, it is possible to predict the response to selection, particularly if a trait is normally distributed and selection is performed by truncation (animals with phenotypes above a certain level). Figure 4.13 illustrates such conditions where $\bar{x}$ is population average and X is point of truncation. The change that can be obtained by selection is known as the response to selection or genetic gain, symbolised by ΔG. The genetic gain by selection is measured as the improvement in progeny phenotypes from selected parents compared to the performance of the previous generation.

The expected genetic gain is:

$$\Delta \mathrm{G} = \mathrm{S} \cdot \mathrm{h}^2 \tag{4.29}$$

where the selection differential (S), measured in the trait unit in question, is the distance between the population average ($\bar{x}$) and the average of selected animals (p). When S is multiplied by the heritability (h^2), which indicates what portion of the selection differential is heritable, an estimate of the genetic gain is obtained.

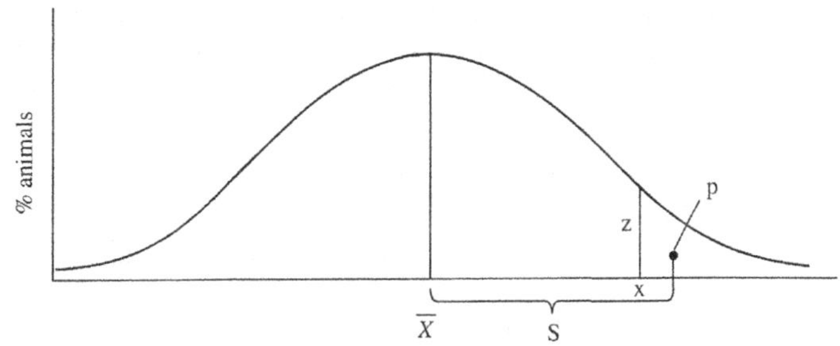

Fig. 4.13 A normally distributed trait with average $\bar{x}$, S is the selection differential, p is the proportion of selected animals, X is point of truncation and Z is height of ordinate at the point of truncation

The selection differential may be standardised by dividing it by σ_P. S/σ_P is called the intensity of selection (i), leading to the following equation:

$$i = S/\sigma_P \quad \text{or} \quad S = i \cdot \sigma_P \tag{4.30}$$

and the expected response to selection becomes:

$$\Delta G = i \cdot \sigma_P \cdot h^2 \tag{4.31}$$

Genetic gain per year can be obtained by dividing ΔG by the length of generation interval (L):

$$\Delta G/\text{year} = i \cdot \sigma_P \cdot h^2/L \tag{4.32}$$

The selection intensity (i) depends on the proportion of animals selected. If selection is by truncation and Z is the height of the ordinate at the point of truncation, then:

$$i = z/p \tag{4.33}$$

If the proportion of selected animals is known and selection is by truncation for a normally distributed trait, then the value of i can be found in Appendix 1.

To illustrate the effect of the different parameters responsible for genetic gain, the expected response to selection is estimated for a range of values using the formula for genetic gain (Table 4.6). All three parameters, selection intensity, standard deviation and heritability of the trait, have considerable effect on genetic gain. Heritability in particular has a very pronounced effect.

The breeder has relatively limited possibilities to influence the parameters of trait standard deviation and heritability, however he has full control over the selection intensity. An increased selection intensity can be achieved through increasing the number of potential breeders. In practice, this means that the testing capacity of the traits in question must also be increased. Heritability can be increased to some

Table 4.6 Response to selection with varying selection intensities, heritabilities and trait standard deviations. Examples are derived from populations of Atlantic salmon with an average weight of 5 kg at harvest

Selection intensity (i)	Heritability (h^2)	Standard deviation (kg)	Genetic gain (kg)	Genetic gain (%)
2.665 (1%)	0.3	1.5	1.2	24
		1.25	1	20
		1	0.8	16
	0.2	1.5	0.8	16
		1.25	0.67	13
		1	0.53	11
	0.1	1.5	0.4	8
		1.25	0.33	7
		1	0.26	5
1.775 (10%)	0.3	1.5	0.8	16
		1.25	0.67	13
		1	0.53	11
	0.2	1.5	0.53	11
		1.25	0.44	9
		1	0.35	7
	0.1	1.5	0.27	5
		1.25	0.22	4
		1	0.18	4
1.400 (20%)	0.3	1.5	0.63	13
		1.25	0.53	11
		1	0.42	8
	0.2	1.5	0.42	8
		1.25	0.35	7
		1	0.28	6
	0.1	1.5	0.21	4
		1.25	0.18	4
		1	0.14	3

extent by standardising the environmental conditions for the animals under test, and taking systematic and accurate records.

The formulas for genetic gain presented above apply to selection for one trait only, and modified formulas and approaches must be implemented when more than one trait is to be selected for simultaneously.

4.9.5 Multiple Trait Selection

In most breeding programs, multiple traits are of interest and hence included in the breeding goal. In aquaculture species, these commonly include traits such as body weight, disease resistance and product quality. There are three basic approaches to multiple trait selection:

- Selecting for one trait in each generation, known as tandem selection
- Setting a threshold for each trait that is used as a threshold of selection, known as the independent culling level
- Selecting simultaneously for all traits considering economic weight, heritability, and phenotypic and genetic correlations between the traits, known as index selection or total score selection.

Hazel and Lush (1942) studied the efficiency of these three methods of selection and concluded that index selection was the most efficient and tandem selection was the least efficient method. Such findings have been subsequently confirmed in independent studies. Index selection is discussed in more detail in Chapter 9.

4.9.6 Correlated Response to Selection

It is well documented that selection for one trait will influence other traits that are genetically correlated. This is known as the correlated response (CR). The response of P_Y when selection is applied to P_X can be used to illustrate the estimation of the size of the correlated response, Fig. 4.8. Following the paths from P_X to P_Y, which are: $h_X - r_{GXGY} - h_Y$, the formula for the correlated response (CR) will be:

$$CR_{PY} = i \cdot h_X \cdot r_{GXGY} \cdot h_Y \cdot \sigma_{PY} \tag{4.34}$$

where i is the selection intensity on P_X, h_X is the path from genotype G_X to phenotype P_X, h_Y is the path from G_Y to P_Y, r_{GXGY} is the genetic correlation between G_X and G_Y and σ_{PY} is the standard deviation of trait P_Y.

When a trait under selection has a negative genetic correlation of a certain magnitude with a trait not under selection, problems may arise since the correlated response may be considerable. To date, there is little evidence of correlated responses in fish and shellfish causing considerable problems. In terrestrial livestock species where intense selection has taken place over many generations, there are several examples of correlated responses causing serious problems. This is discussed in more detail in Chapter 13.

In fish farming, feed costs represent a major proportion of the overall cost of production. Feed conversion ratio (FCR) is therefore a very important economic trait particularly for carnivorous species. However, FCR is difficult and very costly to record in practice. As a result, breeding programs rarely include this trait. Nevertheless, strong genetic correlations have been documented between growth rate and FCR in a number of species, in the order of –0.60 to −0.90 (Gjedrem and Thodesen 2005). The negative correlation implies that a fast growth rate has a low FCR. In rainbow trout, Gjøen et al. (1993) estimated the genetic correlation between growth rate and feed conversion efficiency to be −0.78 and Kinghorn (1981) found a strong positive genetic correlation between growth rate and gross food conversion efficiency (growth/food consumed) in young rainbow trout. These results imply that selection for faster growth rate will also improve feed conversion efficiency. This was documented in Atlantic salmon where FCR was reduced by 25% over five generations of selection for faster growth rate (Thodesen et al. 1999).

4.9.7 Effect of Selection on Genetic Variance

In general, selection will have some effect on genetic variance however it is a relatively complex relationship. According to Bulmer (1971), there is a reduction in the genetic variance particularly after the first generation of selection, known as the Bulmer effect. Therefore, this has led to recommendations that selection should be somewhat less intense in the first generation. Fimland (1979) studied this issue in more detail and conclude that 'The present generalisation of the theory initiated by Bulmer (1971) indicates that any selection force, natural or artificial, with a permanent structure in subsequent generations, leads to a stabilised utilisable state of the additive genetic dispersions. The utilisable additive genetic parameters are those needed for any predictive use in the breeding work'.

The effect of selection on the genetic variance of a quantitative trait depends to a large extent on the number of loci and alleles involved. As the number of genes that affect a given trait increases, the reduction in genetic variance decreases. An underlying reason for this is that as the number of loci increases, the genetic span or range of variation (the distance between the extreme fixed genotypes) also increases. According to Bentsen (1994), the genetic span increases 3.2 and 5.5 times as the number of loci increases from 10 to 100 and from 10 to 300, respectively. This indicates that smaller changes in allele frequency are needed to produce a selection response as the number of genes affecting a trait increases.

After discussing the effect of selection on genetic variation, Lush (1994) concluded that 'Mass selection is generally effective at changing the mean of the population but rarely changes its variation detectably over as short a period as four to five generations'.

However, it should be remembered that the extremely high fecundity of most aquatic species allows very high selection intensity to be practised, potentially leading to a rapid accumulation of inbreeding and a dramatic reduction in genetic variance.

4.9.8 Methods of Selection

There are a number of ways that allele frequencies can be changed within a population:

- Crossing with animals from other populations that show better performance than the starting population, also known as migration
- Selection of breeders with high performance within the population
- Mutation, a continuously process with small and typically harmful effects in the short term

- Genetic drift, a process that occurs by chance and has the most pronounced effects in small populations rather than in large populations.

The first method is of particularly interest when other populations exist that is clearly superior, and is efficient and rapid to implement. Through the use of milt from a superior population to fertilise eggs from the base population, the difference between the two is halved in one generation. Migration is also a strategy to address inbreeding that may have occurred.

Relatively little can be done in practice to reduce the incidence of mutations and genetic drift. If mutated genes result in mortality, then selection will reduce their frequency in the population. The negative effects of genetic drift can be addressed through maintaining a sufficiently large effective population size.

There are several selection methods available for fish and shellfish species:

- Selection based on pedigree information
- Individual or mass selection
- Family based selection, based on measurements on full- and half-sibs
- Selection based on progeny performance

Selection using pedigree information is based on parental records, however since the parents are already selected before mating, pedigree selection has effectively already taken place. Therefore, there will be relatively little new information available for selection in their progeny.

In most fish and shellfish species, many economically important traits cannot be recorded on live individuals. Hence individual selection cannot be performed for these traits, which may include disease resistance and product quality (fat content, fillet yield, flesh colour). The high fecundity of most aquatic species has resulted in family selection being a key strategy in breeding programs for these species. Records from close relatives (typically full- and half-sibs), allows selection for traits like disease resistance and product quality.

In terrestrial livestock species, progeny testing of sires plays an important role in breeding programs, mainly due to the low fertility in females. Since progeny testing substantially lengthens the generation interval, it is not used in aquatic species. However, for species that spawn multiple times, progeny testing may be a useful strategy in a breeding program in order to utilise males with extremely high breeding values. Cryogenic preservation of sperm is another technique that can be used effectively in conjunction with progeny testing.

Selection methods are discussed in more detail in Chapter 7.

4.9.9 Selection Limits

An important issue in selective breeding over the long term is the selection response limits that may potentially be reached. More specifically, do such limits exist? How quickly does genetic improvement reach a plateau? How many generation of selection are possible before this limit is reached? In several selection experiments carried out with laboratory animals, such a plateau was reported after around 30 generations (Falconer 1960). The factors responsible for reaching plateau so early could be a low effective population size (Ne = 15–32), together with relatively few loci affecting the traits under selection.

Other studies however, have documented response to selection over much greater time periods. Enfield (1979) selected for pupa weight in *Tribolium* over 120 generations and found a continuous response to selection. The difference between the selected and control lines was 28 genetic standard deviations at this point, and there was no evidence of a reduction in genetic variation and heritability during this period.

In a selection experiment with mice that lasted for 122 generations, the number of pups borne alive was 22 in the high line (H) compared with 11 in the control line (Fig. 4.14). After 20 generations of selection, the response flatted out, and the high line was subsequently crossed with an unrelated line that had been selected for litter size over 33 generations. Additive genetic variation was constant over the three periods (1–44; 45–70; 71–122) in the high line and the control line, but decreased over periods in the low line. Over the project's duration, inbreeding increased and reached F=0.64 in the control line and F=0.36 in the high line (which was a substantial underestimate). Inbreeding reduced the mean litter size by 0.72 pups per 10% increase in inbreeding. For the high line, this resulted in a reduction of 2.6 pups in generation 122 and if the inbreeding level had been similar to the control population, this would have increased by 4.6 pups.

A long-term selection experiment for fast growth rate in quails was carried out by Marks (1996). One selection line (P) was fed a diet with high protein content (28%) and another (T) was fed with a low protein diet (20%). Selection was performed for 97 generations and the response was highest in the high protein line. Over the period of selection, body weight increased around 3.5 fold in the P line and around 2.5 fold in the T line (Fig. 4.15). An interesting observation was that increased growth rate in the selected lines was accompanied by an increase in feed and water intake, and by an improvement in feed efficiency. No evidence of any major physiological changes accompanying selection for growth rate was observed.

These examples show that selection is an impressively powerful tool to change animal populations in a desired direction. For quantitative traits that are controlled by a large number of genes, selection limits will rarely be reached if inbreeding is kept low.

In discussing the prospects for continued genetic improvement, Hill (2008) concludes as follows: 'Even without a complete understanding of the results from long term selection experiments, it is clear that breeders have been effective in producing very large genetic changes over very long periods, and that there is good reason to expect continued rapid change'.

Bentsen (2005) discussed the possibilities for long term selection response and concluded that: 'By extending the Mendelian laws of inheritance to a simplified model of polygenic inheritance, it is possible to explain how a continuous response

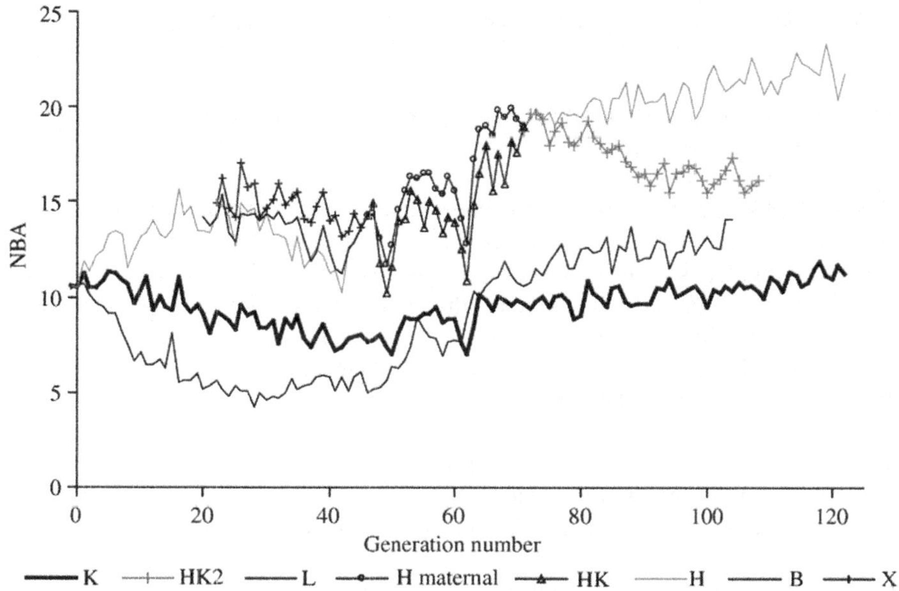

Fig. 4.14 Mean number of pups born alive in the selected lines and the control line. H is the average of line H, B and X. H maternal is the average of the lines H4, H8 and H12. Reproduced from Holt et al. (2005) by permission of Wiley-Blackwell.

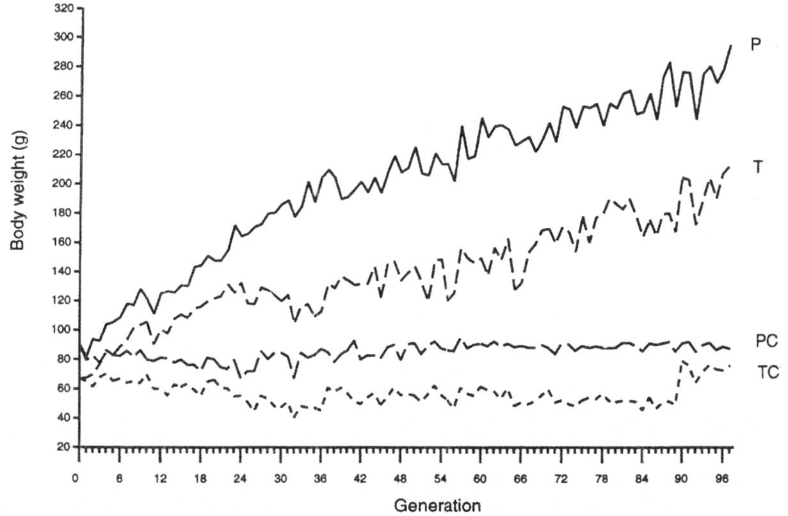

Fig. 4.15 Four-week body weights (g) of four Japanese quail lines selected for body weight over 97 generations (P line on a high 28% protein diet, T line on low 20% protein diet, and their respective controls, PC and TC. Reproduced from Marks (1996) by permission of Poultry Science Association, Inc.

to selection may be observed far beyond the genetic value observed in the base population, and without loss of variation in the genetic effects on the trait. The model considers the simultaneous additive genetic effect of many segregating genes, each with a minor effect on the trait. If the number of genes affecting the trait is sufficiently large, selection limits may be caused by biological constraints rather than fixation of the desired alleles'.

Hill (2008) argued that mutation is one source of fuelling the continuously selection response seen in production traits and also helps to explain why heritabilities are not falling. Cock et al. (2008) also discussed the role of mutations as a source of new genetic variation, and concluded that mutations likely contribute to maintain genetic variation in highly fecund species like *L. vannamei*.

There may be several reasons for reaching a plateau in genetic gain in a breeding program. The most important are:

- Narrow genetic variation in the base population
- Few loci controlling the traits selected for
- Small effective population size that results in inbreeding and increased homozygosity
- Artificial selection may be outweighed by natural selection.

Chapter 5
Initiating Breeding Programs

5.1 Introduction

Wild animals do not necessarily thrive in captivity. In most cases, the confined spaces, unusual surroundings and the general fear of people leads to high stress levels in the animals. This can have detrimental effects on their health and can result in higher susceptibility to disease. Indeed, there are a number of cases where major diseases have only emerged once animals began to be intensively farmed, even though the responsible pathogens had been long present in the wild environment. Farmed fish and other species are usually fed quite different feed to their natural sources of nutrition, often resulting in large amounts of wastage. Once the animals are bred in captivity, natural selection as well as artificial selection will take place and start the process of domestication.

This was precisely the situation when AKVAFORSK initiated research on Atlantic salmon farming in 1971. The wild Atlantic salmon was not suitable for farming, therefore in order to establish a sustainable production system, it was necessary to commence a process of improving welfare and productivity of the fish kept in captive conditions. From the experience gained over many years in terrestrial livestock species, it was known that productive animals could be developed through selective breeding, and in most countries efficient breeding programs were already established for a range of terrestrial livestock and plant species.

This was not the case for aquatic species. No large-scale efficient breeding programs had been implemented in aquaculture at the time. Some selection experiments had shown genetic improvement for diseases. Embody and Hyford (1925) increased survival in brook trout against furunculosis and Schaperclaus (1962) increased the rate of survival considerably in common carp. Kirpichnikov et al. (1972) reported successful selection for improved resistance to dropsy disease in common carp. On the contrary, experiments carried out in Israel in the 1960s involving individual selection for improved growth rate in common carp produced no responses (Moav and Wohlfarth 1973, 1976). These results were well known in the industry, and were used as an argument for the notion that selective breeding simply does not work in fish. As Chapter 3 shows, this position was clearly wrong, and fortunately researchers persisted and succeeded with selection experiments.

T. Gjedrem, M. Baranski, *Selective Breeding in Aquaculture: An Introduction*, Reviews: Methods and Technologies in Fish Biology and Fisheries 10, DOI 10.1007/978-90-481-2773-3_5,

When selective breeding programs in salmonids commenced in earnest around 1970, no estimates of genetic parameters (like heritability) for quantitative economic traits in fish species were available, nor data on genetic and phenotypic correlations among traits. Although basic knowledge about animal breeding was known and several textbooks on breeding theory for farm animals were already published, the application of such methods to aquaculture species required modifications of the standard selection methods developed at that time, primarily to account for the dramatically higher fecundity of fish and shellfish species.

This chapter discusses the different steps involved for developing breeding programs for fish and shellfish.

5.2 The Fundamental Basis of a Breeding Program

The fundamental objective of a breeding program is to lay the basis of sustainable aquaculture production. A productive, domesticated individual will utilise feed, water and land resources far more efficiently than most of the animals currently used in aquaculture productions. For aquaculture industries, the potential for increased production efficiency through systematic genetic improvement is enormous, as genetically improved stock represents less than 10% of the current world aquaculture production.

Breeding programs have long-term goals and objectives. Changes implemented in the current generation are first realised in the next generation. For some species, this lag represents three to four years. Furthermore, typical changes or responses tend to be around 10–15% per generation for many traits, requiring accurate measurements. Experience from the Atlantic salmon breeding program in Norway that commenced in 1975 showed that the farmers who obtained access to smolts from the breeding program did not necessarily see positive results from the first generation of selection, but did so in the second. That meant that in reality, it took eight years from the commencement of selection until the farmers realised the benefits of the breeding program.

To secure necessary capital for investment in a breeding program, investors who are somewhat patient for return of profit are required. It must be clearly explained to investors, however, that there is a very favourable benefit/cost ratio of such investments, mainly because the genetic gain is cumulative over generations.

In the following chapters, the different elements of a breeding program will be discussed. For a breeding program to be successful, it must be planned and implemented with careful attention to detail. This applies to not only the different steps in the practical breeding work such as estimation of population size, testing procedures, estimation of breeding values, selection of broodstock; but also to investments in testing facilities, market assessment for improved product, contracts with test stations and multipliers, and the possibilities for the production expansion for the species in question.

5.3 Establishment of a Base Population

The first, and one of the most important, steps to be taken when commencing a breeding program is the formation of a population with a broad genetic diversity. This will ensure that rapid inbreeding can be avoided and maximise the likelihood of long-term genetic response. Several breeding programs and experiments in fish may have failed in the past because of low genetic variation in the base population (Hulata et al. 1986; Teichert-Coddinggton and Smitterman 1988; Huang and Liao 1990).

There are several alternative ways to establish a good base population:

- If only wild animals are available, broodstock should be selected from at least four genetically diverse strains (Holtsmark et al. 2006)
- Available broodstock of farmed fish with no information of pedigree could be highly inbred, therefore other farmed strains or wild stocks should be included
- If broodstock with known pedigree are available, the level of inbreeding and their effective population sizes should be assessed in order to decide if it is necessary to include broodstock from other farmed or wild populations.

For practical reasons, the first mating after the collection of broodstock from several sources can be random within the strains. However, in the next generation, complete crosses between all strains should be performed in order to form a synthetic base population. This will reduce possible inbreeding at the same time (Holtsmark et al. 2008). It is generally recommended to apply low selection intensity in the initial generations (Eknath et al. 2007), an approach that may secure the maintenance of broad genetic variability for future selection. However, Holtsmark et al. (2007) concluded that prompt, strong selection resulted in greater gain and consistent advantage in the fraction of fixed positive alleles, assuming that the wild populations were unaffected by selection.

There is no single figure for the number of broodstock that should be used to form the base population, but evidence suggests that a minimum of at least 100 males and 100 females should be included. Certainly, even higher numbers would be advantageous if feasible.

Although these basic principles apply to the commencement of a novel breeding program, the actual formation of the base population of present breeding programs in aquaculture species have been quite different, as some examples below show.

For Atlantic salmon in Norway, the collection of broodstock for the base population started in 1971. A total of four base populations were produced, reflecting the four year generation interval of the species. Each of the four base populations included broodstock originating from eight to 24 river strains. From each strain, the aim was to use four males and 12 females to produce 12 full-sib groups and four half-sib groups, however these target numbers were not reached for all river strains (Gjedrem et al. 1991a). The base populations were formed by random mating between and within strains. Progeny from the base population were selected for body weight at harvest both across and within families. The F1 generations for the four populations were produced in the years 1975–1978, and for subsequent generations, selected breeders have been mated randomly under the restriction of not mating closely related animals (full- and half-sibs). Selection intensity for growth rate has been intense from the first generation of selection.

In the GIFT (Genetic Improvement of Farmed Tilapias) breeding program for Nile tilapia in the Philippines, the situation was quite different. The base population was formed from eight unrelated strains consisting of four wild strains from Africa and four strains farmed in the Philippines (Eknath et al. 1993). The first mating was performed within strains while in the second, a complete 8×8 diallel cross was made (Bentsen et al. 1998). The base population was subsequently made up of 25 strain combinations, allowing selection with low intensity to maintain all the strains in the base.

Figure 5.1 shows the contribution of the founder strains to the base population measured as a percentage of the grandparental ancestors. The greatest contributions were from the wild Kenyan and Egyptian strains, while the Thai strain contributed most among the local strains. Eknath et al. (1998) demonstrated that subsequent generations of selection lead to a shift in the representation of the original strains. In the F5 generation, the contribution of the three wild strains from Egypt, Kenya and Senegal increased, while the contribution from the wild Ghanaian strain and all the original domesticated strains decreased.

In India, a selection experiment was carried out at CIFA (Central Institute of Freshwater Aquaculture) with rohu carp (*Labeo rohita*). The origin of the base population was six river strains (Reddy et al. 2002). To form the base population, mating was performed both between and within river strains and the mating was at random.

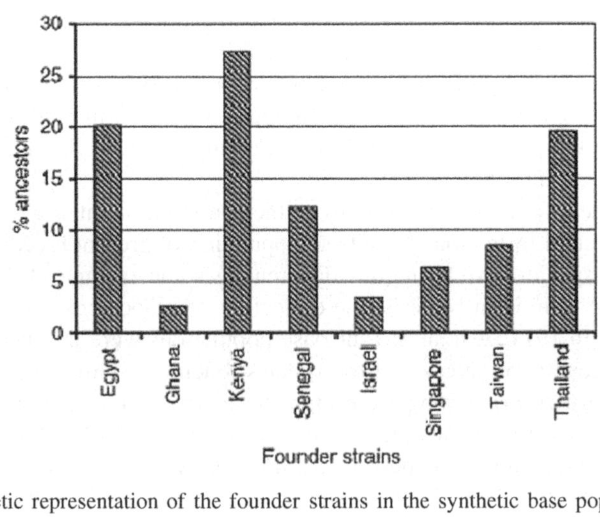

Fig. 5.1 Genetic representation of the founder strains in the synthetic base population of Nile tilapia (*Oreochromis niloticus*) measured by the proportion of purebred grandparent ancestors from each of the founder strains. Reproduced from Eknath et al. (2007) by permission of Elsevier

5.4 Breeding Goal

5.4.1 Introduction

An effective breeding program must have a clear and distinct goal. The overall aim is to develop fish and shellfish stocks with high productivity, or in other words animals with:

- Reduced cost of production (ensuring price competitiveness)
- High product quality (ensuring competitiveness in the market)
- Improved welfare and reduced stress
- Increased resistance to diseases.

Since the sustainability of the industry is paramount, the breeding goal should be broad and to the fullest extent possible include all traits of economic importance. However, it must be kept in mind that as more traits are included in the breeding goal, the genetic gain obtainable for each individual trait will be reduced. Since the effects of selection are realised in future generations, it is important to keep this in mind and therefore look ahead to ensure the best possible breeding goal for future generations.

Breeding companies should establish breeding goals in close collaboration with farmers and consumers, and both processers and exporters should also be consulted. It is essential that all sectors get the possibility to express their wishes to ensure that the best possible products will be produced. However, since the quality demands may vary from one market to another, it may be necessary to define separate breeding goals for different markets. This will often result in an increase in running costs for breeding companies and may even in some cases reduce genetic gain.

Some traits are likely to remain economically important for many generations and will therefore be a long-term part of the breeding goal. For fish and shellfish species, growth rate and survival are examples of such traits. Indeed, it would be highly unlikely to find any aquaculture species where such traits are not of interest to the breeder. The economic importance of some traits may change over time, and some may even change from having a positive value to a negative value! In fish species, the desired values of meat quality traits may change over time and consumer preferences may vary in different markets both within as well as between countries. This creates problems for a breeding program since efforts to improve some quality traits may be a waste of time and resources if the goals are changed. The breeding goals should therefore be adjusted over time to meet the demand from the consumers.

A breeding program may have so called side effects where it causes changes in traits that are not included in the breeding goal. This occurs when genetic correlations exist between traits in the breeding goal, and traits considered to be of less economic importance and traits that were not intended to be changed. In some cases, this might result in negative effects on the correlated traits. To reduce the possibility of such negative side effects, one should systematically record traits that are not included in the breeding goal, but are of importance to overall fitness. Examples of such traits include the rate of fertilisation, occurrence of early mortality, appearance of internal organs, and behavioural traits. If such negative side effects are observed at an early stage, there is sufficient time for the breeding goal to be changed to avoid long-term problems.

Olesen et al. (2000) argue that the development of sustainable production systems also requires consideration of environmental and social concerns, and such priorities should have equal or higher priority than short-term production and economic gains. A good example of such an ethical concern is animal welfare.

There are certain fundamental prerequisites for a trait to be included in a breeding goal:

- The trait must be of economic or ethical importance
- It must show variation and part of the variance must be heritable
- It should be possible to measure the trait accurately at a reasonable cost.

5.4.2 Growth Rate

Growth rate is one of the most important economic traits in all farmed species. Rapid growth has many advantages and is usually crucial for profitability. The economic benefit of faster growth can be realised by a shorter time to harvest, meaning production with a faster turnover rate. Another way to exploit faster growth is to use the original turnover rate, but produce larger animals. A fast growing animal requires less energy and protein for maintenance compared with one that grows slower, and as growth rate increases, so does feed conversion efficiency. The fast growing animals will have a reduced production cost since the maintenance requirements often amount to around 25 percent of the feed cost.

The breeding goal for growth rate is often expressed as body weight at harvest. This trait is easy to record on a scale after the animals have been anesthetised. Growth rate is a long-term breeding goal and will in all likelihood be included in the breeding goal for all aquatic species being farmed today or in the future.

The total phenotypic variation in growth rate has now been studied in a large number of species. The general finding is that the phenotypic variation is large with coefficient of variation (CV) in the order of 20–30%. This figure is very high compared with what is usually found for meat producing livestock species (Table 3.3). Heritability estimates for this trait tend to be in the range 0.20–0.30 (Table 4.1). As discussed in Chapter 3, the response to selection for growth rate has been very high in a number of species. For example, growth rate has been doubled in six to seven generations of selection in Nile tilapia and Atlantic salmon.

Since growth rate is such an economically important trait, it often receives a very large relative weighting in the overall breeding goal. It is therefore necessary to be aware of the potential indirect effects it may have on other traits in the breeding goal and fitness traits.

5.4.3 Feed Conversion Ratio (FCR) and Efficiency (FCE)

Both feed conversion ratio (FCR) and feed conversion efficiency (FCE) are terms used to describe the efficiency that an animal converts feed to growth. FCR is defined as the amount of feed per unit growth (typically kg feed per kg growth). Alternatively, FCE is the measure of growth per unit feed (typically kg growth per kg feed). In carnivorous fish, feed represents up to 60% of production costs while it is substantially less for herbivorous animals farmed in extensive and semi-extensive production systems. For herbivorous fish and crustaceans, the cost of feed will also be high in intensive production systems.

A major problem is how to record FCR in a breeding program. It is not feasible to record FCR for individual animals, since that would require them to be kept in isolation, clearly not normal conditions for fish and shellfish. With the use of special measuring equipment, it is possible to record the amount of feed consumed at a family level in early life stages, but this is an expensive and time consuming task. The optimal time for recording such data would be when animals are larger with a much higher daily feed consumption. However, it is very expensive to rear hundreds of families in separate cages or tanks as they grow larger, and to record the amount of feed consumed and wasted for each. Due to the difficulty and expense of recording this trait, it is not surprising that no breeding programs today are recording this trait.

Since FCR is such an important economic factor in many species, a viable alternative is to use indirect measures of the trait. In terrestrial livestock species, there tends to be a high genetic correlation between growth rate and FCR. For meat producing animals, a genetic correlation of around −0.90 has been reported (Andersen 1977 for cattle; Vangen 1984 for pigs), meaning that as growth rate increases, the FCR will automatically be reduced as a correlated effect. In Atlantic salmon Thodesen et al. (2001) found that feed consumption and feed utilisation may be improved by selective breeding for increased growth. It is, however likely that FCR could be further improved by selection for less energy in weight gain.

Kolstad et al. (2004a) showed that there was considerable variation in feed efficiency (FCE, kg growth/kg feed) among Atlantic salmon families. The study was performed at an early age and in fresh water. The family effect explained 77% of total variation in feed efficiency (Fig. 5.2). Feed efficiency was significantly correlated to growth (0.60) and feed intake (0.45). Kolstad et al. (2005) showed that a cost efficient design can be used to estimate genetic correlations between feed efficiency measured in fresh water and during the grow-out period in sea water, and also between feed efficiency, growth and body composition.

In rainbow trout, Gjøen et al. (1993) estimated a genetic high correlation, $r_G = 0.78$, between feed consumption and growth rate. This is similar to the genetic correlation estimated between growth rate and feed conversion rate in terrestrial livestock species. Therefore, by selecting for fast growth, feed efficiency will be improved through the correlated response.

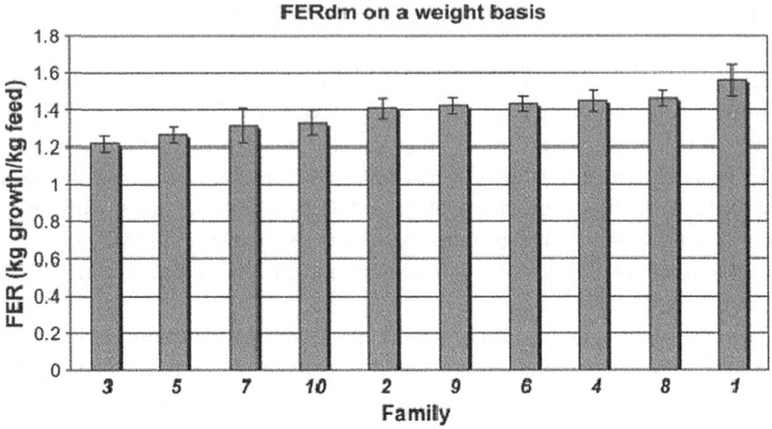

Fig. 5.2 Feed efficiency ratio measured in 10 families from 28 to 40 g live weight. Reproduced from Kolstad et al. (2004a) by permission of Elsevier

5.4.4 Disease Resistance

Diseases create a lot of problems in aquaculture, the major ones being mortality, downgrading (poor product quality) at slaughter, and pollution through the use of antibiotics (Gjedrem 1998). Therefore, aquaculture is frequently named as a risky industry because infectious diseases may cause major problems. Since animals are often kept in waters with high turnover or in lake or sea cages, it is difficult to shelter them against pathogens dispersed in the water. In addition to direct losses of animals caused by disease outbreaks, and the labour required to remove mortalities, there are additional, and often large, costs of medicines and labour to administer them. High survival has, therefore, high priority in the industry and often has a strong focus in a breeding program. Like increased growth rate, disease resistance is considered to be a long-term breeding goal.

Survival is a complex trait and depends heavily on a number of environmental factors that vary during the year and between farming locations. Not unexpectedly therefore, survival as a trait tends to have a low heritability of around $h^2 = 0.00$ to 0.16. (Kanis et al. 1976; Rye et al. 1990; Eknath et al. 1998; Jonasson, et al. 1999; Suarez et al. 1999; Gjerde et al. 2004). Most estimates of survival are based on early life stages. The advent of electronic tags has made it easier to also record mortality on larger animals, but the expected finding of higher heritabilities for this trait in larger animals has not materialised. Therefore, substantial effort has been put into the development of alternative methods to measure disease resistance, and to identify correlated traits that can be used for indirect selection.

Comprehensive investigations have been carried out to study the effects of immunological factors like lysozyme activity, total IgM (immunglobulin), SH-activity (spontaneous haemolytic activity), Anti A-layer (A-protein layer) and Anti O-antigen (O-protein) and their relationship to disease resistance. Lund et al. (1995)

summarised the research conducted up to the mid 1990s and concluded that none of the parameters studied had a genetic correlation higher than ± 0.37 with disease resistance. In addition, the combination of information from several immunological parameters did not result in high correlations with major diseases.

Selection for either high or low stress response in rainbow trout as measured by blood cortisol levels did not show consistent results in challenge tests with different bacterial pathogens. In a furunculosis challenge test, the mortality rate in the high-stress response line was higher than in the low-stress response line, with the opposite being true for a vibriosis challenge (Fevolden et al. 1992). Based on these findings, indirect selection using immunological and stress response parameters has not been applied in breeding programs.

In Table 5.1, some estimates of genetic correlations between growth rate and disease resistance are shown. Most of these estimates consist of data from young fish, however Standal and Gjerde (1987) studied mortality caused by coldwater vibriosis in three year-classes of Atlantic salmon varying from 4.4 to 5.2 kg in body weight. Most estimates presented in the table show positive genetic correlations between growth rate and survival. This implies that selection for growth rate will have a positive correlated effect on survival rate. With genetic correlations around 0.30, the population over time will become more resistant to disease when selection is applied for increased growth rate.

This general conclusion also held for survival of shrimp in ponds and tanks. However, in challenge tests for White Spot Syndrome Virus (WSSV) and to a lesser degree for Taura syndrome virus (TSV), the genetic correlation with growth rate is

Table 5.1 Genetic correlations between different measurements of survival and growth rate

Trait	Genetic correlation	Reference
Survival, fingerlings, brook trout	0.3	Robison and Luemp. (1984)
Coldwater vibriosis., adult, salmon	0.18	Standal and Gjerde (1987)
Survival, fingerlings, Atlantic salmon	0.37	Rye et al. (1990)
Survival, fingerlings, rainbow trout	0.23	Rye et al. (1990)
Furunculosis, challenge, Atlantic salmon	0.30	Gjedrem et al. (1991b)
Survival, fingerlings, Atlantic salmon	0.30	Jonasson (1993)
Fungal infection, Arctic char	0.50	Nilsson (1992)
Survival, fingerlings, Nile tilapia	0.20	Eknath et al. (1998)
VHS, fingerlings, rainbow trout	−0.14 to −0.33	Henryon et al. (2002)
Taura syndrome, challenge, *P. vannamei*	−0.12	Fjalestad et al. (1997)
Survival pond/tank, *P. vannamei*	0.40 to 0.42	Gitterle, et al. (2005)
WSSV, challenge, *P. vannamei*	−0.55 to −0.64	Gitterle, et al. (2005)
No. of sea lice, Atlantic salmon	0.37	Kolstad et al. (2004b)

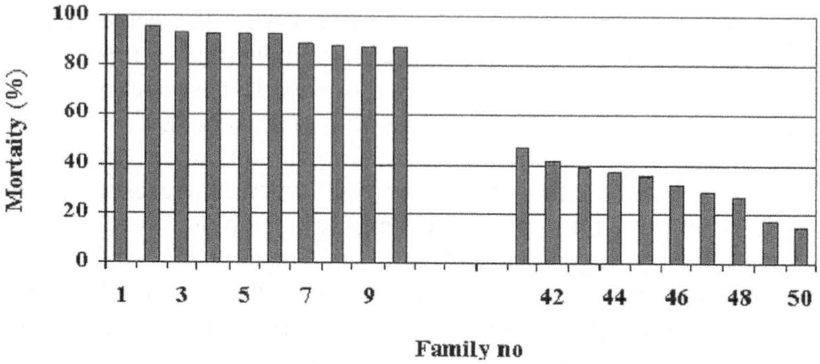

Fig. 5.3 Overall mortality of 50 full-sib families of Atlantic salmon challenge tested with furunculosis. The ranking of the ten families with the highest and lowest mortality during the test is shown

negative, implying that selecting for growth rate increases susceptibility for these viral diseases, particularly for WSSV. Henryon et al. (2002) also reported negative genetic correlation between the viral disease VHS in rainbow trout and growth rate.

The major breakthrough in applying selective breeding for disease resistance was the development of controlled challenge tests. Several experiments have found relatively large amounts of genetic variation in disease resistance by applying such challenge tests. An example is shown in Fig. 5.3 where pre-smolts of Atlantic salmon were tested for resistance to furunculosis bacteria *Aeromonas salmonicida.*

In general, the heritability for each disease recorded in a challenge test is relatively high. Genetic correlations between bacterial diseases in Atlantic salmon are positive but vary in size (Gjedrem and Gjøen 1995; Gjøen et al. 1997). The genetic correlation between bacterial diseases and the virus ISA was found to be low but negative (Gjøen et al. 1997) while Ødegård et al. (2007) found the genetic correlation between furunculosis and ISA to be positive but low ($r_G = 0.15$). Kjøglum et al. (2008) studied resistance to furunculosis, ISA and IPN by applying challenge tests and the estimated heritabilities were 0.62, 0.37 and 0.55, respectively. Genetic correlations between the diseases were low and varied from −0.11 to 0.07.

In addition to bacterial and viral diseases, a number of parasites frequently cause problems for some fish species. In cage culture of salmonids, sea lice often attack the fish, attaching themselves to fins and skin and causing large wounds. Salmon farmers control sea lice infections by chemical or biological means, and chemical treatments are often applied to cover whole regions. A natural biological approach to this problem is the use of cleaner fish or wrasse, that feed on the lice in the first year of production.

Kolstad et al. (2004b) studied the magnitude of genetic variation in the resistance of Atlantic salmon to salmon lice. A challenge test was developed and used to investigate whether salmon families differed in their ability to withstand infection by salmon lice under controlled conditions. The trait was measured as the number

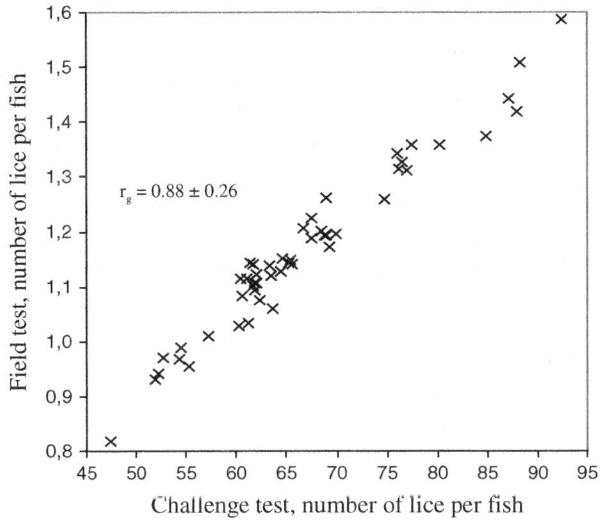

Fig. 5.4 Breeding values for the number of sea lice per fish from 50 Atlantic salmon full-sib families under controlled (challenge test with post-smolt) and natural infection. Reproduced from Gjerde et al. (2007b) by permission of The Research Council of Norway

of sea lice present on an individual, and the heritability estimate was $h^2 = 0.26$. Fish from the same families that were used in the challenge test were subject to a natural infection in sea cages (Gjerde et al. 2007b). The genetic correlation between the field and challenge results was as high as $r_G = 0.88$ (Fig. 5.4).

The most serious disease in farmed Atlantic salmon in Australia is amoebic gill disease. To date, bathing the infected fish in freshwater has been the only effective treatment, however this is a costly and time-consuming exercise. The results of a challenge test of 30 full-sib families indicated a broad sense heritability of 0.16 for gross gill score and 0.35 for image gill score (Taylor et al. 2007). The authors concluded that the findings provide good scope for selective breeding to increase resistance against amoebic gill disease.

An accidental introduction of the monogenean ectoparasite, *Gyrodactylus salaris* into Norwegian salmon rivers in the mid-1970s has decimated 46 Norwegian river populations. To study the genetic variation and heritability, a controlled challenge test was performed (Salte et al. 2009). In total 973 fingerlings of about 8 g from 49 full-sib families were infected by cohabitation. Eleven families were wiped out, 15 families had between 10 and 25% survival and the four least affected families had survival rates between 36 and 48%. Estimates of heritability of survival on the liability scale was 0.32 and time until death for fish that died showed also considerable genetic variation with a heritability of 0.29. The authors concluded that selection of survivors as parents for the next generation is expected to almost double the overall rate of survival in one generation.

Vertebral deformities recognised as individuals with anomalous body shape are regularly observed in farmed fish. Many studies have been performed to study the causes of skeletal deformities in cultured fish species. The established causes include genetic, environmental, nutritional, infectious and toxic factors. The incident of vertebral deformities in Atlantic salmon was observed on a total of 44,684 progeny of 225 sires and 471 dams (Gjerde et al. 2005). The deformities were classified as humpback or a shortened tail of ungutted fish at different farms across four year-classes. The large variation between the year-classes in deformities (9.5, 7.6, 21.5 and 2.3%) could not be explained by differences in water temperature during egg incubation. The genetic correlation between deformity and body weight were negative indicating that that high genetic growth potential was not the cause of deformity, and there was no evidence to suggest that inbreeding caused the deformities. The estimated heritabilities on the liability scale for the different year-classes were 0.36, 0.22, 0.25 and 0.00. The estimate of zero heritability was partly explained by the low incidence of deformities that year of 2.3%. Gjerde et al. (2005) recommended to avoid selecting breeders from families with a high incidence of deformed fish and to completely avoid selecting breeders with any signs of deformities.

5.4.5 Age at Sexual Maturation

The traits discussed so far are likely to be an important part of the breeding goal for nearly all aquatic species, however this is not the case for the trait age of sexual maturation. In aquaculture production, it is undesirable for animals to become sexually mature before they reach market size. The main reason for this is that the production of gonads is a very energy demanding activity, often resulting in a substantial reduction in growth rate (growth can cease altogether!), a reduction in product quality, and a potential increase in mortality during the maturation process. Therefore, the breeding goal for some species is to reduce the frequency of animals maturing before they reach market size. Some species naturally meet this requirements (carp, catfish, milkfish), while others reach sexual maturity too early (tilapia, Atlantic cod, Atlantic salmon).

As shown in Table 4.1, age at sexual maturation is a heritable trait of a medium magnitude. There are relatively few and varying estimates of phenotypic and genetic correlations between age at sexual maturation and other economic traits. Gjerde et al. (1994) found a positive genetic correlation between body weight until 16 months in sea water and age of sexual maturity in Atlantic salmon ($r_G = 0.11$ to 0.49). Similar results were obtained by Gjerde and Gjedrem (1984) for Atlantic salmon ($r_G = 0.49$ to 0.52) and for rainbow trout ($r_G = 0.11$). These results indicate an unfavourable relationship between these two traits, implying that selection for faster growth rate will increase the frequency of early sexual maturation.

Some males of Atlantic salmon become sexual mature before smoltification (as parr or precocious males). At one year of age they are around 30 percent smaller compared with non-mature fingerlings (Gjerde 1984). In the sea, fish that had matured as parr had a similar growth performance as non-parr. Gjerde (1984) concluded that maturation as parr is a heritable trait in Atlantic salmon and that it is independently inherited from maturation in sea.

5.4.6 Product Quality

Product quality is for most species a relatively unstable trait over time because consumers frequently change their preferences for what defines the best quality. The reasons for these changes in consumers preferences are very complex and unpredictable. New findings on the effect of nutrition on human health, as well as research on the nutritional benefits of consuming fish and shellfish will change the market demand for different species as well as increasing demand for particularly beneficial quality phenotypes. This situation will lead to a change in the breeding goal for quality traits.

Some external characters may be important for species sold whole. The generally desirable appearance can be described as a normal form with typical proportions, a muscular body and skin colour common to the species. For these traits, there is a lack of suitable methods for accurately measuring them and they are therefore often measured with a subjective score that usually has a low heritability. For internal traits like fat content and distribution, fillet colour and texture, it has not been possible to take measurements on living animals and therefore individual selection has not been possible. There is strong potential for increased genetic gains for these traits if instrumentation could be developed to measure quality traits on live animals. Alternatively, these traits make good candidates for the approach of molecular marker-assisted selection. Folkestad et al. (2008) documented promising methods for measuring fillet colour and fat content on live Atlantic salmon with high predictability.

There are a number of quality traits of relevance to aquaculture species, with the relative importance varying depending on the species in question. A selection of these are described below.

Carcass size: For most species, the market is willing to pay more for particular sizes of fish and shellfish, which has consequences for when animals are harvested. For the farmers, this is a problem because there always will be a large amount of variation in the size of fish and shellfish in a pond or a cage at harvest time even though they are of the same age. It is possible to grade the fish and harvest animals of a certain size, but this is laborious and causes stress to the animals. This stress prior to slaughter has been implicated in having a negative effect on product quality.

This is a universal problem in polyculture, where different species are reared in the same unit and must be separated at harvest. In addition, all species will not reach market size at the same time and therefore some must be returned to the pond/cage to continue to grow. This necessitates an increase in the handling of animals which is both laborious and causes stress.

Fillet yield: Fillet yield is defined as the consumable part of the body and is the ratio between fillet weight and body weight. This trait is considered to be one of the most important economic quality traits in most species. Fillet yield is expensive to record, particularly when fish are marketed as whole fish. At present, this trait cannot be recorded directly on live animals and selection must therefore be based on family averages. Atlantic salmon has a very high fillet yield percentage (Table 5.2). For both rainbow trout and Atlantic salmon, fillet yield is a relatively highly heritable trait (Table 4.1).

Table 5.2 Yield of Atlantic salmon carcass. Reproduced from Kittelsen et al. (2002) by permission of Gan forlag

Part of fish carcass	Percent
Whole fish	100
Gutted carcass	85
Gutted without head	75
Fillet without backbone	69
Fillet without backbone and earbone	61
Fillet without backbone, earbone and pinbone	60

Fillet colour: For salmonid species, it is essential that the fillet is of a particular shade of red, not too pale and not too dark. In the wild, salmonid flesh obtains its red colour from astaxantin, a carotenoid found in the crustaceans that form part of the feed of the species. Farmed fish gain their red coloration through feed in form of artificially produced astaxanthin. Added astaxanthin is expensive and represents around 15% of the feed cost. Only a small portion of astaxanthin is retained in fish of less than one kilogram in size, around 4%, while it can be up to 32% in larger fish (Bjerkeng 2008). Heritability of fillet colour is medium to high in Atlantic salmon, with one estimate of, $h^2 = 0.47$, (Table 4.1).

The specific breeding goal for increasing colour in salmonids is actually to increase retention of astaxanthin in the fillet. A major problem has simply been to accurately measure the colour in the fillet. It can be measured by chemical analysis, however this is expensive and time-consuming. There are no methods in widespread use today for measuring colour in whole carcasses or living animals.

For fish species with white flesh, like tilapia, carp, catfish, and cod, discoloured fillets with grey and or yellow spots are considered to be of poorer quality.

Fat content: Fillet fat content is an economically important trait for salmonids and several other species with high fat content. The consumer's preference varies a lot between and within markets and therefore it is difficult to define the optimum fat percentage. Generally, the smoking industry tends to prefer a fat content of between 15 and 17% in the fillet of salmonid species. In addition, the processing industry prefers little or no variation in fat percentage. From a breeding perspective, it is very difficult to select for reduced variation in a trait.

For species with a low fat percentage like Atlantic cod, Atlantic halibut, sea bream, grey mullet, shrimp and clams (Haard 1992), fat content will not be part of the breeding goal, however it has been found to have a high heritability (Table 4.1)

Fat distribution: Fish tend to store fat in depots around the body, however the location of these depots differs between species. Salmonids store fat in the belly, around fins and intestine, Atlantic halibut have fat depots around the fins and cod have depots in the liver. These fat depots reduce product quality, increase the cost of production and represent waste in processing.

Texture: Fillet texture is the characteristic of the fillet that can be felt by hand and when consumed. It is generally measured in terms of hardness and juiciness.

Negative consumer reaction is most pronounced when the fillet is too soft. Texture is of importance both in fresh and processed products. To date, it has not typically been included in breeding goals although it is an economically important trait and has been shown to be heritable ($h^2 = 0.26$) (Table 4.1).

Intramuscular bones: Some fish species like carp, silver barb and sea bream have intramuscular bones. These small y-shaped bones are found along the lateral line and are difficult to remove during processing. Some studies have documented genetic variation for the number of intramuscular bones in common carp (Segenbusch and Meske 1967) while others did not find genetic variation (Moav and Finkel 1975).

Other quality traits: Dressing percentage is the proportion of the body minus the amount removed at gutting (intestine, blood, fat in and around intestine) divided by body weight. Figure 5.5 shows that there are large differences between species in dressing percentage. While Atlantic salmon has a high dressing percentage, it is low in Atlantic cod. For larger fish, the differences are even more pronounced. Cod would have a larger liver and rainbow trout would have more fat around the intestine. It is a clear goal to increase the dressing percentage, however to do so introduces some complications. In Table 5.2, the different parts of the body of an Atlantic salmon are illustrated. The problem of including dressing percentage in the breeding goal is that the intestine is a major component of this trait, and a major change in the size of the intestine could have negative effects on the metabolism of the animals. It would be advantageous however, to reduce the size of the head and in particular the fat depots around the intestine.

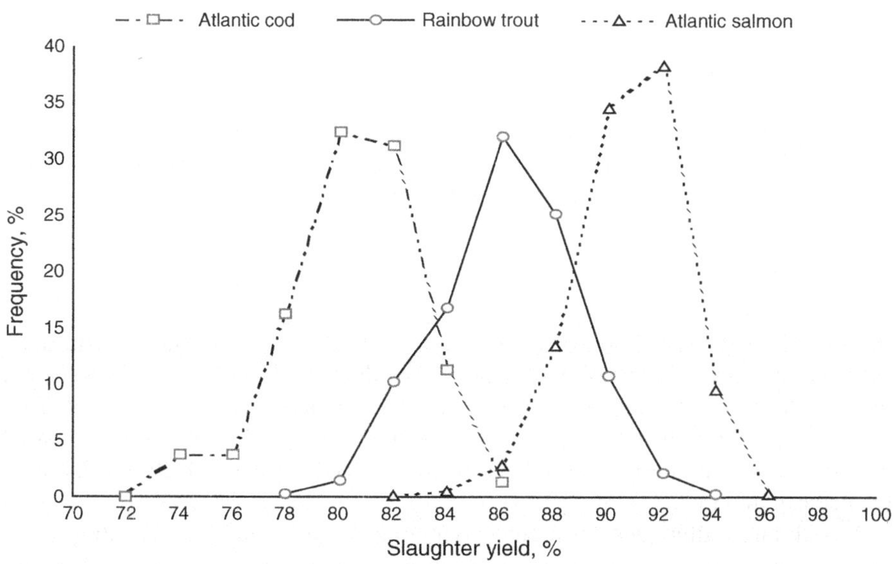

Fig. 5.5 Dressing percentages for different farmed fish species. Reproduced from Rørå et al. (2001) by permission of Wiley-Blackwell

Taste and smell are another two important quality traits. At present, these traits are only measured by test panels, which is very laborious and expensive. It is therefore not possible to include them in a breeding goal. Gaping is another quality trait which can not be measured by instruments, but can be judged by visual scoring.

5.4.7 Cold Tolerance

Tolerance to low temperature is a problem for tropical species when they are farmed in temperate and some subtropical regions. Their culture is highly affected by sensitivity to low temperature, leading to poor growth and mass mortality during overwintering. Tilapia originating from tropical and subtropical parts of Africa are now farmed throughout the world and commonly face this problem.

A challenge test for cold tolerance can be applied by lowering the water temperature for fingerlings until death occurs. Charo-Karisa et al. (2005) challenged fingerlings of Nile tilapia using such a test. Mortality of fish started at 13.6°C and total mortality occurred at 8.6°C. Heritability for temperature at death was 0.09 and for cooling degree hours 0.08.

Kolstad et al. (2007) studied methodology for analysing survival data in fish. They found low heritabilities for cold water tolerance in Nile tilapia from a challenge test in Vietnam. Highest heritability was found for time until death, h^2 = 0.06 – 0.07 while survival until 50 or 80% of the population was dead showed very low heritability.

5.4.8 Fecundity

Fecundity is an extremely important reproduction trait in all terrestrial livestock species, since it tends to be very low. In most fish and shellfish species however, it is extremely high and is therefore no candidate for selection. It is, however, necessary to record fecundity and survival in early life stages in order to observe if the traits are negatively affected by selection on other traits.

5.4.9 Behaviour

In aquaculture, behaviour plays a central role in production and management. In an early phase of the domestication of a species, wildness will be obvious. Containment must be strong, and the high stress levels that the animals experience means that they tend to perform poorly at this stage. They are more susceptible to diseases compared with domesticated animals. This often results in feed wastage which will increase feed conversion rate.

It is an interesting question if behaviour could be included in the breeding goal? The main problem is how to record the trait. At present, there are no available methods for recording behaviour, wildness or tameness of individuals or families. Until more is known about the trait behaviour, domestication alone must be relied upon to address this issue. It has been observed that Atlantic salmon selected for seven to eight generations are much calmer and less fearful of humans compared to wild fish.

5.4.10 Recapture Frequency

For anadromous species that reproduce in fresh water and have their grow-out phase in the sea, sea ranching is an interesting option. It is largely practised with the salmon species in the Pacific. Dr. Lauran Donaldson performed pioneering work in sea ranching when he released smolts of Chinook salmon from a small pond on the University of Washington campus in Seattle and had them successful return to the pond in 1953 (Donaldson 1968). The most economically important trait in sea ranching is the frequency of fish returning to the point of release, or recapture frequency. This is a heritable trait (Carlin 1969) with a heritability of $h^2 = 0.08$ (Jonasson et al. 1997).

Figure 5.6 shows a relatively large amount of variation in recapture frequency between families of Atlantic salmon. Jonasson (1994) reported a response to one generation of selection for recapture frequency of 27%. It should therefore be possible to increase recapture frequency by applying family selection.

5.4.11 Central Breeding Goals

The breeding goal must be individually defined for each species because the economically important traits differ between species and the marketing situations may vary in different countries. However, the most important traits will most likely be:

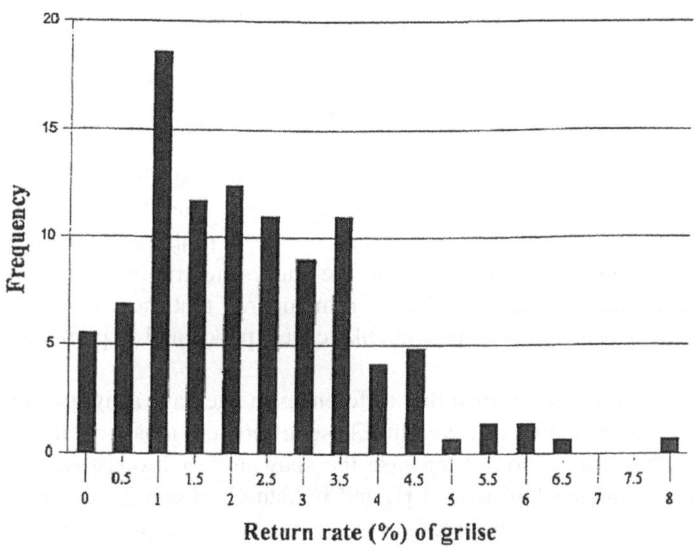

Fig. 5.6 Frequency distribution of grilse (one year at sea) return rate from 145 full-sib groups tested in the 1991 year-class. Reproduced from Jonasson et al. (1997) by permission of Elsevier

In aquaculture farming:

- Growth rate
- Disease resistance
- Quality traits
- Age of sexual maturation (for some species).

In sea ranching:

- Recapture frequency
- Growth rate.

5.5 Registration of Records

5.5.1 Introduction

Traits included in the breeding goal must be clearly and precisely defined. It must be emphasised when and how the trait should be measured and recorded. All personnel involved must be instructed and trained in how the recording should be performed. The animals should be handled with care in order to stress them as little as possible. Before measurements are taken, they should be anaesthetised and calm, since this increases the likelihood of obtaining unbiased records. The importance of taking accurate measurements cannot be over-emphasised, as it is vital to avoid error both in recording and in taking notes of the records. Although all efforts are made to take accurate records, sometimes extreme, unrealistic records will appear in the data. These records are also known as outliers or errors. Before analysing the data further, these outliers must be deleted from the data because they provide false information on the individual in question and for their family. Outliers will also affect estimates of variance.

As pointed out in Section 4.5.5 it is important to keep in mind that the individuals as well as the families and progeny groups must be treated and reared in such a way that environmental differences will be minimised. Communal rearing of families and progeny groups is the most effective way to reduce these environmental effects. It should also be remembered that there may be relatively large differences between rearing units like cages, ponds and in particular between farms.

Repeatedly it has been shown that differences in age have a significant effect on several traits, in particular body weight. These differences must be reduced as much as possible. One way is to synchronise the spawning of broodstock by applying mammalian luteinizing hormone, LH, and human chorionic gonadotropin (HCG) which has proven effective in inducing maturation and ovulation in fish. Since it is not possible to obtain the same age of all families, body weight must be corrected for age differences before breeding values are estimated.

5.5.2 Body Weight

As arguably the most economically important trait, growth rate should be recorded at harvest and measured accurately. In addition, weights should be recorded during grow-out in order to follow the development of the stock. Whether the animals are reared in ponds or cages, they must be gathered before the weights are recorded. This process must be done with care to ensure the animals are stressed as little as possible. One way to achieve this is with a soft dip net to catch the individuals for anaesthetising before they are put onto the scale.

5.5.3 Survival

Disease causes major problems in aquaculture production and must be reduced or avoided completely if possible. For bacterial diseases, vaccines have been developed for several diseases and have in general shown excellent results in preventing disease outbreaks.

It is obvious that wild animals as well as animals early in the domestication process are stressed in captivity and it is well known that this stress increases susceptibility to diseases. Farmers have a number of ways to minimise stress, including reducing noise and light intensity (dusk), moderating density in the tank/pond/cage, frequently feeding to reach all animals, optimising oxygen levels with moderate water current, and reducing traffic of people and machines.

In general, survival under rearing conditions shows very low heritability, $h^2 = 0.04 - 0.16$ (Table 4.1). This implies that response to selection will be low. Therefore, Gjedrem (1995) proposed to use standardised challenge tests to study and record genetic variation in disease resistance for the most serious diseases. The high fertility of aquatic species makes this possible, while the low fertility of terrestrial livestock species means such strategies cannot be used. Challenge test protocols are now available for several diseases and species. In simple terms, the procedure involves sampling and tagging around 20–30 individuals from each family, followed by placing them in a communal tank. After the test animals are acclimatised, a batch of animals from the same population that were earlier infected by a parasite, bacteria or virus, are put into the tank to pass on the infection to the test animals. This procedure is called infection with cohabitation. The advantage of this type of challenge test is that all defence mechanisms of the animals will be operating (for example mucus, skin, and intestine) For some viral diseases, cohabitation tests are difficult to perform and therefore each animal must be injected individually with the virus. This bypasses the outer defence mechanisms of the animal, and so this type of challenge is not as 'realistic' as the cohabitant challenge.

As the animals start to die, the individuals are removed, identified, and time of death recorded. In some cases, the challenge tests are terminated at 50% overall mortality while in others mortality continues to be recorded until it ceases. Results from the challenge test are used to rank the families for resistance against the disease.

At present, it is not possible to use the survivors from viral disease challenges as broodstock because they may be a carrier of the diseases they have been challenged with. In some cases, broodstock from bacterial challenge tests have been used after treatment with antibiotics, but this is not a widespread practice. A possible solution could be to grow the survivors from the challenge test in isolation and at maturation strip the males and use the milt for fertilisation of selected eggs. Alternatively, other methods such as marker-assisted selection could be used to select individuals within families.

For economic reasons, only small animals are challenged with diseases. This raises the question of what is the genetic correlation between survival in a challenge test and resistance to the disease under farming conditions. Gjøen et al. (1997) found a very high genetic correlation ($r_G = 0.95$) between a challenge test for furunculosis of 20–40 g fingerlings of Atlantic salmon in freshwater and a field test after 2½ months in marine cages. Similar results were obtained by Ødegård et al. (2006) where the estimated genetic correlation between a field test and challenge test for furunculosis was found to be $r_G = 0.71–0.75$. Kolstad et al. (2004b) found a strong genetic correlation of ($r_G = 0.80$) between the challenge test and natural attachment of lice in sea cages (Fig. 5.4).

The Norwegian Atlantic salmon breeding companies routinely carry out challenge tests for the bacterial disease furunculosis and the viral diseases: ISA (infectious salmon anaemia) and IPN (infectious pancreatic necrosis). More recently, controlled challenge tests have been introduced in breeding programs for rohu carp (testing for resistance to *Aeromonas hydrophila*), and for resistance to TSV and WSSV in penaeid shrimp. Likewise challenge tests are currently under implementation for selection in tilapia.

5.5.4 Feed Conversion Efficiency

As discussed earlier, feed conversion efficiency is not directly recorded in breeding programs for direct selection because it is so expansive and difficult to do so. However, the trait can be part of the breeding goal and included in a selection index with its economic value, heritability, genetic and environmental correlations with other traits in the breeding goal. The use of a selection index will enable the use of all the parameters to maximise the genetic improvement for all the included traits. The selection response will to a large degree rely on the magnitude of the genetic correlations with the other traits in the breeding goal. The strong and favourable genetic correlation between FCR and body weight will be a key factor for obtaining a positive genetic gain for this important trait.

5.5.5 Product Quality

From a breeding perspective, the main difficulty with product quality traits are that they are difficult to record on live animals and that the traits vary with age and size of the animals. To further complicate the matter, the market may change over time because of changing consumer preferences.

For fish with high levels of fat, like salmonids, the optimum fat level tends to be one of intermediate magnitude. In practice, it is not possible to use chemical analysis to measure fat content in fillets as it is too time consuming and expensive. Computerised tomography is a non-destructive method which is highly accurate in predicting fat content of fillets (Rye 1991; Gjerde 1987). The amount of fat depots can also be estimated by computerised tomography. Technology for image analysis has been developed with a high degree of accuracy to predict fat content in fillets for salmonids ($R^2 = 0.83$). Since there are so many advantages with measuring the trait on living animals, technology such as that introduced by Folkestad et al. (2008) would be of great advantage to the industry. Experiment trials using such near infrared spectroscopy (NIR) equipment resulted in a correlation between NIR measurements on live salmon and fat content of $r = 0.94$.

Flesh colour is an important quality trait in salmonids. The carotenoids that produce this colour must be given to the fish in the feed, which is an expensive task and the retention rates are relatively low. Subjective score has been used to measure the red colour intensity of fillets but this form of colour measurement has a low heritability (as is similar for other traits measured by subjective scoring). Rye et al. (1994) used a colorimeter reading device (Minolta Chroma Meter CR-300) and obtained a high heritability using such measurements. The new methods introduced by Folkestad et al. (2008) for measuring fillet colour on living animals is a promising development for enabling within family selection for this trait. The correlation between visible spectroscopy predicted on gutted salmon and chemically measured pigment was $r = 0.85$ (Folkestad et al. 2008).

Flesh texture is a quality trait of economic importance. Generally, fillets should have firm, cohesive and elastic properties. Several instruments have been developed to measure texture, including a texture analyser, TA-XT2. (Stable Micro System). This instrument utilises a flat-ended cylinder probe (diameter: 12.5 mm) equipped with a load cell of 5 kg. The force needed to penetrate the fillet surface in called the breaking strength. Mørkøre and Einen (2002) found that sensory hardness was highly correlated ($r = 0.70$) with a Warner-Bratzler blade of 12.5 mm in diameter in raw salmon fillets. Mørkøre and Rørvik (2001) showed also that this instrument effectively quantified the variation in texture, which varies during the year. They found a negative relationship between breaking strength and specific growth rate. However, no instruments are available to measure texture on live animals.

Taste is another important quality trait, however there are no instruments available to measure it in a practical manner. An organoleptic judgement is the only alternative but it is very costly to perform.

5.6 Adjustment of Data

In Fig. 4.6, the environmental factors influencing animals are divided in two parts: systematic and random. In a selection program, the environmental variation should be kept as low as possible in order to increase the heritability and consequently genetic gain.

Table 5.3 Average body weight (kg), standard deviation (σ) and CV for female, male and immature rainbow trout and Atlantic salmon

	Rainbow trout			Atlantic salmon		
Sexes	Body weight	σ	CV	Body weight	Σ	CV
Female	3.71	0.85	23	4.75	1.09	23
Male	4.16	0.95	23	5.75	1.38	24
Immature fish	3.08	0.98	32	3.61	1.26	35

The random environmental factors can be reduced by standardising the testing conditions as much as possible. Families should be reared communally. If it is not possible to rear all families in one unit, the families should be divided so that all families are represented in each rearing unit. Finally, data should be adjusted for possible differences between units. In addition, it is possible to reduce the random environmental effects by repeated measurements of animals.

Systematic environmental effects can be reduced by estimating correction factors and adjusting the data. Such correction factors should be estimated on a large data set and they should be re-estimated over time. In Table 5.3, the effect of sex on body weight is shown. These effects can to some degree be eliminated by adjustment. Using the data in the table as an example, adjustment can be performed by subtracting 0.45 (4.16–3.71= 0.45) from the weight of all male rainbow trout to give them the same average as females, and by adding 0.63 (3.71–3.08 = 0.63) to all sexually immature fish. This results in three groups with the same average. The difference in standard deviation between the sexes creates an additional problem because adding or subtracting does not affect the standard deviation. If two groups have a different CV, a multiplicative correction will give the groups equal averages and also a similar standard deviation. Using the data for rainbow trout in Table 5.3, a multiplicative correction of 0.89 (3.71/4.16 = 0.89) will adjust the weights of male fish to become similar to the weights of females with equal averages and with similar standard deviations.

Typically, a population of fish or shellfish will include individuals of different ages, which will affect the traits included in the breeding goal. These variations should also be eliminated by adjustment. The most common correction factor used for age differences is the coefficient of regression of age on weight. Usually, the coefficient of regression is positive which means that older animals are heavier than younger, and by adjustment, weight of older animals than average will be reduced and the weight of animals younger than average will be increased.

Under realistic farming conditions, there are large differences between farms and even between cages or ponds within farms. These differences are most likely to be environmental and must be taken into consideration when selecting breeders. If fish in two farms have equal breeding values, one can select equal numbers from each farm as broodstock or an adjustment may be applied to animals on one farm to obtain the same average and standard deviation as the other.

Adjusting records of different traits for systematic environmental variation will reduce the total phenotypic variation, while the genetic variation will be unchanged, as illustrated in Fig. 4.7. In this way, the proportion of genetic variation will be increased, the value of heritability will be higher, and the estimates of breeding values will be more accurate.

However, when standard MME (mixed model equations) analysis is applied, the program will simultaneously fit fixed systematic effects including covariates and random additive effects. This negates the need to make prior adjustments to the data.

Ultimately, one of the most important factors is the level of training and care taken when recording traits and handling animals. All the calculations performed subsequently will depend on the accuracy of these measurements, therefore care and expertise at this stage will ensure more reliable and meaningful results for the breeder to utilise.

Chapter 6
Breeding Strategies

6.1 Introduction

The selection of parents for subsequent generations can be performed in many ways, in most cases the method of choice is strongly influenced by the reproductive biology of the species and desired breeding goal. Alternative selection methods can broadly be grouped into three strategies; inbreeding, crossbreeding and purebreeding. The fundamental difference between these strategies is the degree of relationship between the animals that are mated, and the utilisation of different components of genetic variance. While purebreeding primarily exploits additive genetic effects, inbreeding and crossbreeding strategies exploit dominance effects and epistasis.

6.2 Inbreeding

As discussed in Chapter 4, inbreeding generally leads to detrimental effects on fitness, survival, growth rate, high frequency of deformities, and reduction in genetic variance. Hence breeding programs in livestock and aquaculture species tend to avoid inbreeding, or at least minimise it to an acceptable level. However, when breeding was focused on the formation of new breeds or strains, inbreeding was actively exploited as a strategy. Desired qualities of a breed tend to be a uniform size, body form, colour and colour pattern. Inbreeding helps to achieve such a goal through the increase in homozygosity and reduction in genetic variation.

With the aim of increasing productivity through exploiting additive genetic variance, the maintenance of genetic variation is vital. In this context, inbreeding is a problem and must be avoided as much as possible. A well documented pedigree is a key tool to avoid inbreeding, however in a closed population, inbreeding will inevitably accumulate over time. As a general rule, increases of 0.5% or less per generation are desirable, and up to 1% per generation tolerable.

For some purposes, particularly in laboratory experiments, it may be of interest to use highly inbred lines that are genetically stable (Komen 1990). If non-additive genetic variance (dominance and epistasis) constitute a major part of the genetic variance for important economic traits, inbreeding may be used to develop lines for

T. Gjedrem, M. Baranski, *Selective Breeding in Aquaculture: An Introduction*,
Reviews: Methods and Technologies in Fish Biology and Fisheries 10,
DOI 10.1007/978-90-481-2773-3_6,

subsequent crossbreeding. To utilise this strategy, a breeding company must develop at least two inbred lines that can be crossed to produce hybrids for on-growing. The inbred lines will generally have a relatively low performance, and in order to ensure that the resulting hybrids are competitive with purebred animals offered by programs that apply selective breeding, the inbred lines used must be continuously improved through selection.

If a breeding program does not control the accumulation of inbreeding in the target population, inbreeding may rapidly reach high levels and counteract further response to selection.

6.3 Crossbreeding

Crossbreeding is defined as the mating of animals from different species, strains or inbred lines. The objective is to obtain offspring expressing hybrid vigour or heterosis. From a breeding perspective, crossbreeding is the opposite of inbreeding; inbreeding increases homozygosity whilst crossbreeding increases heterozygsity. When heterosis occurs, the offspring surpasses the average of its parents for one or more traits. Heterosis may also be defined as a superior average performance of offspring compared to the best parental strain. In aquaculture farming, the latter definition is preferred.

In reviewing the status of hybridisation between salmonid species, Chevassus (1979) concluded that in most cases the hybrids showed intermediate, or at best, equal growth to that of the superior parent. This is in agreement with the finding of Refstie and Gjedrem (1975) and Refstie (1983a). Though not widely practised, cross-species hybridisation has been applied in Australia where blacklip abalone (*Haliotis rubra*) and greenlip abalone (*Haliotis laevigata*) hybrids are commercially produced. Based on the results from crossing different river strains of rohu carp, Reddy et al. (2003) concluded that crossbreeding would be of little interest to a breeding program for rohu carp (Fig. 6.1). A more extensive discussion of heterosis effects reported in aquaculture species is presented in Chapter 4.

In terrestrial livestock species, inbred line crosses are frequently used, particularly in poultry breeding programs. To investigate the feasibility of such a strategy in salmonids, AKVAFORSK carried out an experiment with rainbow trout. The results indicated a heterosis effect of the same magnitude as the inbreeding depression for both growth and survival. Gjerde (1988) concluded that the cost and time delay in developing and test-crossing inbred lines would only be justified with larger heterosis effects in crossbred stocks than what was found in this study. Furthermore, Falconer and Mackay (1996) concluded that, in the absence of selection, inbreeding followed by line crossing in a large population would not be expected to make any permanent change to the population mean. According to their finding, the heterosis effect will be halved in the second generation (F_2) when random mating is performed among animals from the first cross (F_1).

Figure 6.2 illustrates that the effect of one generation of selection with genetic gain of 10% will be equal to crossbreeding strains with 10% heterosis. If heterosis

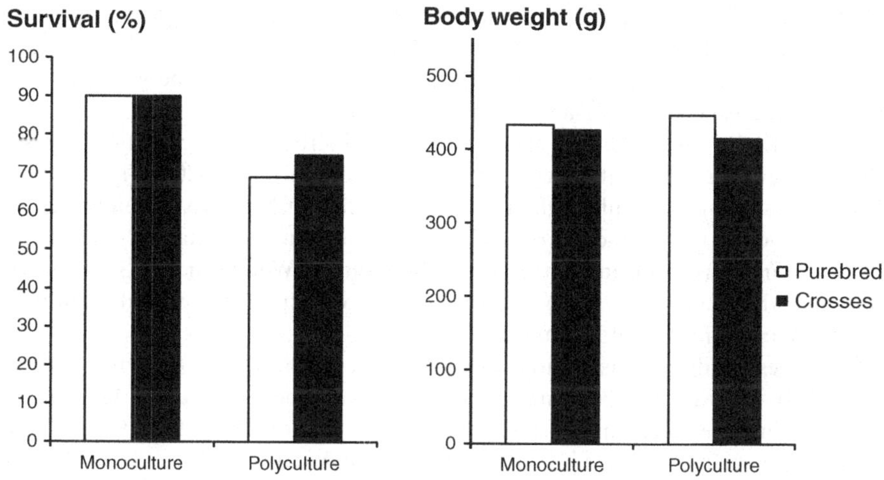

Fig. 6.1 Survival from tagging to harvest and harvest body weight of purebreds and crosses of rohu carp in monoculture and polyculture production systems. Reproduced from Reddy et al. (2003) by permission of Central Institute of Freshwater Aquaculture, India

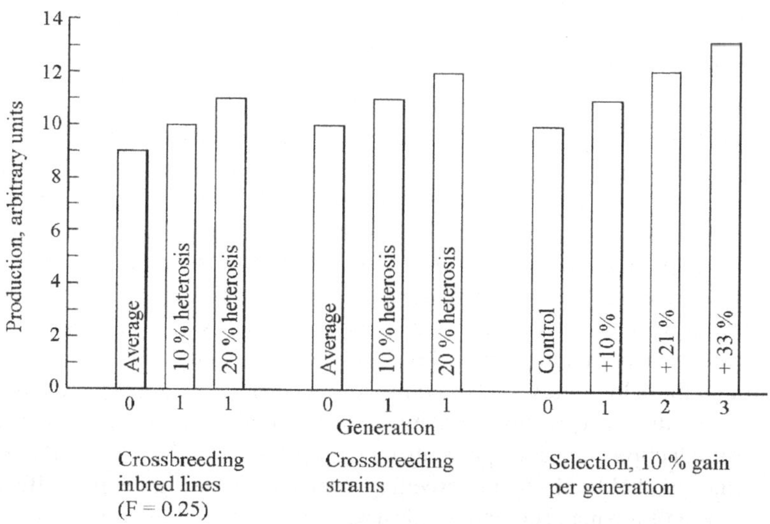

Fig. 6.2 Efficiency of crossing inbred lines and strains with medium and high heterosis compared with relatively low genetic gain from selection. Reproduced from Gjerdem (1985) by permission of Springer

is 20%, which is a level rarely seen in crossbreeding experiments, it will take two generations of selection to reach the same level as crossbreeding with no reduction of heterosis. Crossing inbred lines will lag behind because one generation of crossbreeding is needed to balance the inbreeding depression.

These results highlight the unpredictability of crossbreeding, with heterosis effects ranging from slightly negative to positive gains of 15–20%. This unpredictability is mainly a result of the fact that the effect of heterosis depends mainly on dominance or the interaction of alleles within loci, meaning that only particular gene combinations result in hybrid vigour. In addition, Wohlfarth (1993) state that crossbred advantage appears to be largely limited to relatively young and small fish, with a reduced effect on old animals.

From these findings, it is obvious that long term selection will outperform crossbreeding if the trait has a certain level of additive genetic variance. An alternative is to combine selection and crossbreeding. This is of particular interest if the non-additive genetic variance is considerable. A combined selection and crossbreeding strategy called reciprocal recurrent selection (RRS) was developed by Comstock et al. (1949) and Dickerson (1952). This design, described in detail by Falconer and Mackay (1996), is quite complicated and involves making many crosses between two or more strains or inbred lines. The progeny groups are subsequently ranked and the sires and dams showing the best combining ability are selected to be used to produce the crosses in the next generation. RRS programs are used by commercial poultry breeders and have produced promising results in corn. However, direct comparison with other selection methods has not been encouraging (Calhoon and Bohren 1974), and RRS can only be used for multiple spawners.

If the non-additive genetic variance is considerable, a more simple combination of selection and crossbreeding can be applied, utilising at least two lines with different origin. For each line, testing and selection can be performed in order to obtain the maximum genetic gain per line. This will require double testing capacity compared to an ordinary selection program dealing with only one population. The two lines can be used to produce crossbred animals for in-house production or for sale to the industry. Such a strategy can provide a means of maintaining the security of genetically improved material. Breeding companies in the poultry and pig industries have used a strategy of producing inbred lines to be subsequently used for crossbreeding. Through the sale of only crossbred progeny, it is possible to keep the pure lines in-house. The advantage of this breeding strategy is that the crossbreds are not suitable for reproduction and hence prevents unauthorised use of the improved material. Parental lines are held only by the breeding company, allowing them to effectively control the use of the material in the industry.

6.4 Purebreeding

Mating of unrelated animals within the same population is known as purebreeding. In practice, this means that the relationship between the animals that are mated is approximately the same as the average relationship between animals in the population. Purebreeding is regarded as the method of choice for long term, continuous genetic improvement. Through selection of superior phenotypes, superior gene alleles will be passed on to the next generation, and animals carrying such alleles will have high breeding values. As it is not possible to measure the true breeding value of an individual, breeding value estimates represent an approximation based on the observed phenotypes.

Purebreeding is relatively easy to perform, and is particularly effective when a strain is identified to be equal to or better than alternative strains. Inadvertent inbreeding is a serious risk when applying the purebreeding strategy, and must be minimised as much as possible through avoiding the mating of close relatives like full-sibs, half-sibs and cousins. Sufficient numbers of broodstock should be used in each generation, because inbreeding will accumulate rapidly if the effective population size is low. If individual pedigrees are unknown, then an even larger number of broodstock is required to avoid the mating of close relatives (Bentsen and Olesen 2002).

As illustrated in Fig. 6.2, genetic gain from selection using a purebreeding strategy compares very favourably with crossbreeding. Even with a high heterosis level (20%), it will only take two generations of selection with a relatively low genetic gain of 10% per generation, to produce the same increase in production. The problem with crossbreeding is that the improvement in form of heterosis is maximal after one generation, and if no selection is applied in the lines, there will be no further improvement. Applying continues selection in a purebreeding scenario leads the cumulative genetic response over many subsequent generations, (see Fig. 2.1).

Chapter 7
Selection Methods

7.1 Introduction

Within the framework of the general breeding methods described in the previous chapter, there are several selection methods that can be applied with the common goal of improving the productivity of the animals. A fundamental result of selection is the alteration of allele frequencies in the population. This is particularly evident with genes with additive effects. However, the actual method of selection will have a substantial impact on the extent of these changes in allele frequency. Changes in allele frequencies in the population can also occur through natural biological processes. This chapter describes the processes that lead to changes in allele frequencies in a breeding population, and the various selection methods that can be applied given the particular circumstances unique to each breeding program.

7.2 Factors Affecting Allele Frequencies

7.2.1 Migration

Migration is a method of particular interest when a given population is clearly inferior to other populations. Through the introduction of broodstock from superior populations, it is possible to obtain a rapid genetic improvement. For example, the magnitude of the improvement that can be achieved through fertilising eggs of the target population with milt from a superior population will be halved in one generation.

7.2.2 Selection

Selection is the most important strategy to make long-term changes to a population, and is particularly the case for species with high fecundity. A high fecundity allows for high selection intensity to be practised, a strategy that typically results in a very large response to selection (see Chapter 3).

T. Gjedrem, M. Baranski, *Selective Breeding in Aquaculture: An Introduction*, Reviews: Methods and Technologies in Fish Biology and Fisheries 10, DOI 10.1007/978-90-481-2773-3_7, © Springer Science+Business Media B.V. 2009

7.2.3 Mutation

Mutations occur naturally and at a fairly stable rate in living organisms, often resulting from something going wrong during the cell division process. Most mutations are repaired immediately by the cells themselves, and therefore cause no ill-effects to the animal. Mutations in genes are commonly recessive in nature, and therefore their harmful effects are particularly evident in inbred animals that have a higher likelihood of being homozygous for such genes. This is a key reason to avoid inbreeding as much as possible in a breeding program.

7.2.4 Genetic Drift

Genetic drift is the process of change in allele frequencies that occurs entirely by chance, and is an important concept in population genetics. These random events can change the makeup of the gene pool slightly, are compounded over time, and can ultimately determine which alleles will be carried forward while others disappear. Genetic drift may have significant effects in small wild populations, but is generally not important in controlled breeding programs with a large effective population size.

7.3 Choice of Selection Method

The different selection methods that can be applied in a breeding program were introduced in Chapter 4. For aquatic species, individual and family based selection are the most commonly used strategies. In terrestrial livestock species, pedigree selection and progeny testing are usually practised because females tend to have very low fecundity and produce only a few eggs per ovulation.

The overall aim of all selection schemes is to maximise the probability of correctly ranking all potential breeders with regard to their breeding value. This is essentially the same concept as maximising the correlation between the true and estimated breeding value (r_{HI}). An animal's breeding value can be defined as the average performance of an infinite number of its progeny, or from a practical point of view its ability to produce good or bad progeny.

The choice of selection method for a given situation (species, production environment, scale) depends on a range of factors including:

- Target traits for which genetic improvement is desired
- Feasibility of recording the trait on live animals
- Magnitude of heritability for the traits in question
- Reproduction capacity of the species.

Each selection methods that will be described in the following have advantages and disadvantages that will influence their choice given a particular breeding

Table 7.1 Maximum theoretical value of the correlation between true and estimated breeding values (r_{HI}) for different methods of selection with varying heritabilities

Selection methods	In general	Heritability (h^2) 0.1	0.30	0.50
Pedigree	$\frac{1}{2}h^2$	0.22	0.39	0.50
Individual	$\frac{1}{2}r_{HI}$	0.32	0.55	0.71
Family:				
• half-sibs	$\frac{1}{4}r_{HI}$	0.50	0.50	0.50
• full-sibs	$\frac{1}{2}r_{HI}$	0.71	0.71	0.71
Within family:				
• half-sibs	$\frac{1}{4}h^2$	0.16	0.27	0.35
• full-sibs	$\frac{1}{2}h^2$	0.22	0.39	0.50
Combined selection(1)		0.73	0.77	0.82
Progeny testing		1.0	1.0	1.0

(1)Combined family and within family selection

Reproduced from Gjerde (1991) by permission of SIAS Boksmia.

scenario. However, for most methods, there is a theoretical maximum value of the correlation between the true and estimated breeding values, given different heritabilities for the traits in question (Table 7.1).

7.4 Pedigree Selection

Pedigree selection uses information from the parents and grandparents of the animals in question. In a breeding program where selection is being applied, ancestors are already selected before mating and therefore pedigree selection has already taken place. Pedigree selection is of most interest for young animals without data on their own performance. For these animals, the best estimate of their breeding value is the average breeding value of their parents. The accuracy of pedigree selection is relatively low since despite the fact that progeny inherit half of their genetic material from each parent, Mendelian segregation will cause variation in the breeding value among their progeny. This low accuracy, coupled with the general availability of information from family members, means that pedigree selection is of less importance in aquatic species.

7.5 Individual Selection

Individual selection is based on the performance of each individual and is also known as mass selection. Individual selection is easy to perform and was for many years the most commonly used method of selection in aquatic species. There are many examples of large genetic gains obtained in breeding programs based on individual selection.

However, individual selection is only possible for traits that can be measured or recorded on live animals, since live individuals are naturally required as broodstock for the next generation. In practice, individual selection has only been implemented on a large scale in fish and shellfish species for morphometric traits like body weight and length. Whilst this approach has been successful in many fish species, it is an imprecise approach in shellfish where shell weight often comprises the largest proportion of overall weight. The accuracy of individual selection largely depends on the heritability of the target trait (see Table 7.1). High heritability implies that a large portion of the trait variation is heritable and that the accuracy of individual selection is high. For traits with low heritability, the response to selection will be low because environmental factors explain a large portion of the variation and mask the genetic component.

Technological advances and new methods may facilitate individual selection for traits that earlier could not be measured on live animals. An example is the recent development of near infrared spectroscopy (NIR) methods to measure fat content and pigment concentration in live Atlantic salmon with high accuracy (Folkestad et al. 2008).

Individual selection does also have a number of shortcomings. The efficiency of selection relies to a large degree on environmental conditions being standardised for all animals during their whole life span. If animals have been held in different ponds, tanks or cages, there will be considerable variation in these environmental conditions. One solution is to statistically adjust for differences between different grow-out units. If the animals are not tagged, it is not possible to adjust for systematic environmental factors like differences in age and sex, and such conditions will reduce the accuracy of selection.

A breeding program selecting for growth rate only will usually not require physical tagging of breeding candidates. Since growth rate has a relatively high heritability, there is a significant risk that a large proportion of selected breeders will be close relatives when intense selection is practised. The result will be increased inbreeding. As time passes, genetic gain will be reduced and the viability of animals will decrease. It is likely that over the history of aquaculture, many breeding programs have failed because they used strong selection in each generation without pedigree information and suffered a rapid accumulation of inbreeding and ultimately dramatically reduced productivity. These problems can be largely reduced by using a large number of broodstock in each generation, as demonstrated by Bentsen and Olesen (2002) and discussed in Chapter 12.

7.6 Family Selection

Full-sibs will on average have half of their gene alleles in common, while half-sibs will share one quarter of their identical gene alleles. This relationship implies that the performance of siblings can be used as a basis for selection. As a result of the large family sizes that are commonly seen in aquatic species, the accuracy of breeding value estimation is high and can reach $r_{HI} = 0.71$ for full-sibs and 0.50 for half-sibs independent of heritability values (Table 7.1) Family information therefore has great value for estimating breeding values for aquatic animals.

Family selection is of particular importance for traits with low heritability, such as general survival and age at sexual maturity. The efficiency of family selection rests on the fact that the phenotypic deviations of individual animals as a result of environmental effects tend to cancel each other out in the mean value of the family. Therefore, the phenotypic mean of the family is a good measure of its genotypic mean, and the advantage gained is greater when environmental deviations constitute a large part of the phenotypic variance. Hence traits of low heritability make excellent candidates for the family selection approach, and a low accuracy in measuring individuals will be compensated for by the information obtained by family members.

For traits that cannot be currently measured on live individuals like product quality traits, family information is essential. Recording these traits on sibs makes it possible to estimate breeding values with high accuracy. Disease resistance is another excellent candidate for family selection. Under farming conditions, it is not possible to record disease on individuals, however mortality can be estimated for each family if animals are tagged and reared together. Heritability estimates of this type of data tend to be very low (see Table 4.1), however, when applying controlled challenge tests for specific diseases, heritability estimates are usually of medium to high magnitude. Family selection is also much more efficient than individual selection for threshold traits like age of sexual maturation, particularly when the incidence of the trait is low.

To apply family selection, it is necessary to know the parentage of each individual and hence important to maintain good pedigree records. This usually necessitates individual tagging of animals. Since it is not possible to physically tag the animals at hatching, each full-sib family must therefore be reared in separate units from egg stage through to the commencement of feeding until they reach sufficient size to be physically tagged. During this period, each family will have a common environment that is different from other families. This common environmental effect will to some extent reduce the accuracy of the use of the family average as a prediction of genetic merit. Therefore, this period should be made as short as possible. Refstie and Steine (1978) estimated the common environmental effect to account for about 10% of body weight at smolt stage in Atlantic salmon, however the effect was found to be insignificant at harvest time when the families had been reared communally in sea cages during the grow-out phase (Gunnes and Gjedrem 1978). This stresses the importance of providing all families with as equal environmental conditions as possible during the testing period in order to minimise the common environmental effects.

An alternative to physical tagging is to use genetic markers such as microsatellites to establish the pedigree among the animals tested. This allows families to be reared communally during their whole lifespan and therefore eliminates common environmental effects. This method is becoming more practical as genotyping costs continue to fall, however unequal survival of different families during communal rearing at early stages can bias results. Some physical tagging will, however, continue to be required at time of multi-trait recording and selection of broodstock.

The genetic components estimated from full-sibs contain an additive (σ_A) and a dominance genetic component (σ_D). For paternal half-sibs, the estimated genetic component contains only additive genetic variation. The accuracy of breeding values are less for half-sibs ($r_{HI} = 0.50$) compared with full-sibs ($r_{HI} = 0.71$). Since only the additive genetic variance is transmitted to offspring in a predictable way, it is important to produce half-sibs in addition to full-sibs. Earlier it was shown that heritability estimations should be based on additive genetic variance. Therefore, in a breeding program where family selection is being applied, both full- and half-sibs must be produced.

7.7 Within-Family Selection

When within-family selection is applied, families are tested in separate units and selection is based on the deviation of each individual from its family average. The family average is ignored in this selection strategy. This method is of particular interest when the common environmental effect is large, since selection within-families will eliminate this common environmental effect.

Within-family selection requires testing facilities for each family until they reach market size. There is no need for tagging and it is easy to reduce inbreeding through avoiding the mating of related individuals. Like is the case with individual selection, it is not possible to apply within-family selection for traits that cannot be measured on live animals.

According to the results presented in Table 7.1, the accuracy of within-family selection is much lower than family selection. Gall and Huang (1988a) compared expected selection responses from different selection methods and concluded that combined selection is expected to produce a response per generation about 10–30% above that of individual and family selection and about twice that expected for within-family selection.

7.8 Progeny Testing

Progeny testing as a means of selection exploits the fact that the relationship between each of the parents and its progeny is 50%. The average performance of progeny groups of a certain sire or dam will be a good expression of its breeding value for a given trait, since the progeny group represents a random collection of the respective parent's gene alleles for that trait. If the number of progeny is large, the accuracy of the breeding value will approach 1.0 (Table 7.1). Progeny testing of sires is a central selection method in terrestrial livestock mainly because of low fertility in dams. In aquatic species, progeny testing may be applied for both sires and dams.

Progeny testing has some of the same advantages as family selection because it can be used to select for traits that cannot be measured on live breeding candidates like disease resistance and product quality. However, the big disadvantage of progeny testing is that it will extend the generation interval considerably. Indeed, in many cases progeny testing implies a doubling of the generation interval. This is an important consideration since the annual genetic gain is a key measure for the efficiency of the program. For species spawning only once or having a high mortality rate after spawning, progeny testing is of no relevance. Fundamentally, this is the reason why progeny testing is not an important selection method in aquaculture and is therefore rarely used.

Gall and Huang (1988b) compared different methods of selection and concluded that selection of males based on progeny testing is expected to produce the largest response per generation. However, the long generation interval required to complete each cycle of selection would result in annual genetic gains less than expected for sib selection.

Nevertheless, progeny testing may be of some interest for multiple spawning species. Elite sires or dams may be mainly used for dissemination purposes. In addition, repeated matings can be used to create genetic ties across year classes and/or different batches of families, as will be discussed in connection with estimation of genetic gain in Chapter 11.

7.9 Correlated Response and Indirect Selection

As discussed earlier, some traits in the breeding goal may be very difficult and expensive to record, like feed conversion ratio (FCR). Since this is such an important trait economically, a simpler direct or indirect measure of this trait may be of great value. For fish, a strong genetic correlation between FCR and growth rate has been documented (see Section 5.4.3). This means that selection for increased growth rate will automatically lead to improved FCR. This indirect effect on FCR is called a correlated response, the magnitude of which can be expressed by equation 4.3.4.

For the FCR/growth rate situation, it is known that the genetic correlation between the traits is strong, selection intensity for growth rate can be very high, and heritabilities for both traits are of medium size. Accordingly, the potential for a correlated response in FCR is good. This was demonstrated by Thodesen et al. (1999), who estimated a reduction of 20% in FCR after five generations of selection in Atlantic salmon. The first two generations of selection were performed only for growth rate, and in the following three generations, combined selection was performed for fast growth rate and reduced early sexual maturation. This example shows that the effect of a correlated response can be quite high.

Some traits in the breeding goal may be genetically correlated with traits of minor economic importance that are not targeted for direct selection. The latter traits could still be important for the fitness and welfare of the animals. If such genetic correlations are negative, correlated responses may be harmful to the animals in the long term. In poultry and pigs, several negative effects have been observed as a result of correlated responses, particularly concerning leg weaknesses (discussed in Chapter 13). With this in mind, a suitable precaution to take is to carefully monitor traits that could potentially be negatively affected by selection. Examples are egg number, mortality in early stages, and abnormalities in body shape and organs. This monitoring should facilitate detection of negative effects of selection at early stages, before serious damage has taken place. If a negative correlation to a fitness trait is apparent, the selection procedures may need to be changed.

Likewise, there may also be traits of no direct economic value with high genetic correlation to traits in the breeding goal. Such traits can be used for indirect selection.

A study at AKVAFORSK and Institute of Veterinary College in Norway can serve as an example. Previous studies had showed that mortality in fish had low heritability and the possibility for genetic improvement was therefore weak. This raised the question of whether immunological parameters could be used as an indirect measure of disease resistance. Several immunological parameters were analysed on individual fish and the results were correlated with mortality of relatives which were challenged for several diseases. It was, however, concluded that indirect selection for disease resistance would not be efficient because the genetic correlations with immunological parameters were weak and all below $r_G = 0.31$ (Lund et al. 1995).

7.10 Combined Selection

Combined selection is a strategy that combines all available information in order to maximise the accuracy of the estimated breeding values. The data used may be recorded on the breeding candidate itself, from full- and half-sibs, progeny and parents. Combining information from all of these sources in an optimal way will give the highest possible accuracy of breeding value estimation.

The selection methods of most relevance for aquatic species are individual, family and within-family selection. Table 7.2 presents the theoretical accuracy of breeding values estimated for different family structures with both family and individual phenotypic information available. As expected, the accuracy increases as the number of full-sib members increases, but there is little to be gained by increasing the number of progeny per full-sib group beyond 50. Likewise there is relatively little to be gained in accuracy when each male (or female) is mated to more than three females (or males) in a nested mating design for production of half-sib groups. In fact, the accuracy for two half-sib groups is quite good compared with three, and this is particularly the case when the environmental effects are insignificant.

The accuracy of combined family and individual information increases with higher heritabilities, however, the accuracy of breeding values is strongly reduced as the environmental effect common to full-sibs (c^2) increases. This highlights the importance of standardising the environmental conditions during the testing period, and communal rearing is the best way to achieve this.

Table 7.2 Correlation between the true and estimated breeding value (r_{HI}) for different family structures of half-sib and full-sib groups. Values in parenthesis represent the correlation when individual phenotypes are also available. h^2 is the heritability and c^2 is the common environmental effect for full-sibs

			Number of animals per full-sib group, n			
h^2	c^2	Number of females mated to each male	10	20	50	100
0.10	0.00	1	0.42 (0.48)	0.51 (0.56)	0.60 (0.64)	0.65 (0.67)
		2	0.44 (0.50)	0.52 (0.57)	0.61(0.64)	0.65 (0.68)
		3	0.45 (0.51)	0.53(0.58)	0.61 (0.64)	0.65 (068)
		5	0.48 (0.53)	0.54 (0.59)	0.62 (0.65)	0.65 (0.68)
		10	0.50 (0.55)	0.56 (0.60)	0.62 (0.65)	0.66 (0.68)
0.25	0.00	1	0.54 (0.65)	0.61 (0.69)	0.66 (0.73)	0.68 (0.74)
		2	0.56 (0.66)	0.62 (0.69)	0.66 (0.73)	0.68 (0.74)
		3	0.56 (0.66)	0.62 (0.70)	0.67 (0.73)	0.68 (0.74)
		5	0.57 (0.67)	0.62 (0.70)	0.67 (0.73)	0.69 (0.74)
		10	0.58 (0.67)	0.62 (0.70)	0.67 (0.73)	0.69 (0.74)
0.10	0.10	1	0.33 (0.42)	0.36 (0.45)	0.39 (0.47)	0.40 (0.48)
		2	0.35 (0.44)	0.38 (0.46)	0.41 (0.49)	0.42 (0.49)
		3	0.37 (0.45)	0.40 (0.48)	0.43 (0.50)	0.44 (0.51)
		5	0.40 (0.47)	0.43 (0.50)	0.46 (0.52)	0.46 (0.52)
		10	0.44 (0.50)	0.46 (0.52)	0.48 (0.54)	0.49 (0.55)
0.25	0.10	1	0.45 (0.60)	0.49 (0.62)	0.51 (0.64)	0.52 (0.64)
		2	0.47 (0.61)	0.50 (0.63)	0.52 (0.64)	0.63 (0.65)
		3	0.49 0.62)	0.52 (0.64)	0.54 (0.64)	0.54 (0.65)
		5	0.51 (0.63)	0.53 (0.64)	0.55 (0.66)	0.55 (0.66)
		10	0.53 (0.64)	0.55 (0.65)	0.56 (0.66)	0.57 (0.66)

Reproduced from Gjerde (1991) by permission of SIAS boksmia.

7.11 Multiple Trait Selection and Index Selection

A breeding program will usually have several traits in the breeding goal. Hazel and Lush discussed the principles around selection for multiple traits as early as in 1942, and compared three methods of multi-trait selection. The most efficient method was to select simultaneously for all traits giving each trait a particular economic weight and taking into account the heritability and phenotypic and genetic correlations among the traits. This method is known as index selection. The second most efficient method utilises independent culling levels where a thresholed level is set for each trait which represent the culling level for that trait. The least efficient method of selection for multiple traits was tandem selection, which focuses selection on one trait at the time until a desired genetic level for that particular trait is reached, before targeting the second trait, and so on.

Hazel (1943) was the first to outline the principles for construction of selection indexes. The basic requirements for constructing a selection index are that unbiased estimates of phenotypic and genetic parameters are available, as well as economic weights for each trait in the breeding goal.

Two types of selection indexes are of particular interest for fish and shellfish:

- Indexes for individual animals
- Indexes for families.

Family selection indexes are very useful for pre-selection of potential broodstock while individual indexes are used for the final selection. The indexes include all available information about individuals, full- and half-sibs, and economic values of each trait. A selection index (I) for individuals may have the following form:

$$I = b_1(X_1 - \bar{x}_1) + b_2(X_2 - \bar{x}_2) + \ldots + b_i(X_i - \bar{x}_i) \quad (7.1)$$

where X_i are records for different traits in the breeding goal with an average of $\bar{x}_i$, and b_i is a weighting factor for each trait depending on its:

- economic value
- heritability
- variation
- phenotypic and genetic correlations with other traits.

Indexes of full-sib families will be of similar form and include information of all family members and all half-sibs.

Separate selection indexes must be estimated for each species, and each breeding company will generally develop their own selection indexes. These indexes are complicated to estimate as they have many factors, and require extensive knowledge of statistics to be accurately calculated. The statistical method most commonly used for both terrestrial livestock animals and aquatic species is known as BLUP (Best Linear Unbiased Predictor). Gjerde (2005a) describes the development of selection indexes for aquatic species in greater detail.

By using an index for the selection of broodstock, their progeny will be animals with the best economic value for the farmers. The magnitude of the genetic gain in a breeding program will directly be influenced by the accuracy of the selection indexes used. To supply breeding companies with new knowledge for improving selection indexes is an important objective for research institutions.

7.12 Comparing Different Selection Methods

It is useful to compare the efficiency of different selection methods. Falconer and Mackay (1996) compared the effect of individual, family and within-family selection with the efficiency of combined family and within-family selection. The results are illustrated in Fig. 7.1. The correlation between phenotypic values of family members is known as the intraclass correlation (t) and is estimated as:

$$t = r \cdot h^2 + c^2 \quad (7.2)$$

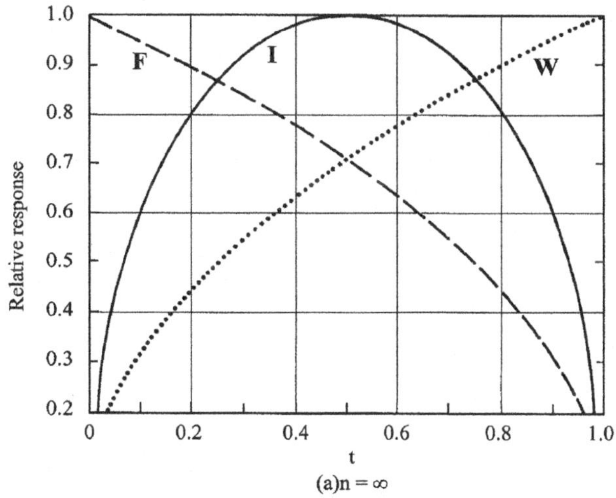

Fig. 7.1 Relative merits of the different methods of selection. Responses relative to that for combined selection plotted against the phenotypic intraclass correlation, t. I = individual selection; F = family selection; W = within-family selection. Reproduced from Falconer and Mackay (1996) by permission of Pearson Education Inc.

where r is the correlation between family members, h^2 is heritability of the trait in question and c^2 is the proportion of the total variance accounted for by common environmental effects.

When the intraclass correlation (t) varies from around 0.25 to around 0.75, individual selection is more efficient than both family and within-family selection. This is because individual selection exploits the entire scope of genetic variation present, while family and within-family selection only utilises part of the genetic variation.

When the intraclass correlation is low, family selection is more efficient than individual selection. Within-family selection compares favourably to both family and individual selection when the intraclass correlation is very high. Individual selection is approximately as efficient as combined selection when t = 0.50, and for such an interclass correlation, family selection and within-family selection are equally efficient.

When the intraclass correlation is high due to members of a family being more alike phenotypically rather than genetically (t > r = 0.50), within family selection is more efficient than both individual and family selection. However, this is a very rare scenario and the primary advantage of within-family selection is the reduction of common environmental effects between families (Uriwan and Doyle 1986).

A comparison of expected selection responses from individual, family, within-family and combined family and within-family selection demonstrated the superiority of the combined selection strategy (Gall and Huang 1988a), a finding supported by the theoretical response shown in Fig. 7.1.

Chapter 8
Mating Design

8.1 Introduction

In Chapter 7, different methods of selection were described. The next step in a breeding program is to decide on the particular mating design that best suits the species, production environment and breeding goal of interest. The simplest breeding program design used in practice is the application of individual selection for growth rate only. Mass spawning can be utilised in such a design, but to prevent rapid accumulation of inbreeding it is advisable to practise artificial mating. Well-designed breeding programs usually include several traits in the breeding goal and therefore combined family and within-family selection is the selection method of choice. Another important aspect of a sustainable breeding program is the identification of males and females and recording of pedigree information of all broodstock used each generation. In this way it is possible to fully control the rate of inbreeding. To produce full- and half-sibs, artificial mating is the preferred method. However, if artificial fertilisation is difficult to implement, natural spawning of broodstock pairs held in separate tanks or hapas may be used to produce families.

A number of different mating designs are used for aquatic species and the most common are discussed in the following sections.

8.2 Mass Spawning

Mass spawning describes the practice of having a number of males and females in a tank or pond and allowing them to reproduce randomly. This is a simple and cheap method that has been common practice in aquaculture for many years. Depending on the spawning behaviour of the species, a range of methods are used to collect the fertilized eggs or larvae for subsequent transfer to hatching or grow-out units.

Brown et al. (2005) studied mass spawning in gilthead sea bream. They found high variance in family size and a large number of non-contributing broodstock, especially males, which reduced the effective population size (Ne) markedly. Another investigation of mass spawning of sea bream was performed in Italy and

T. Gjedrem, M. Baranski, *Selective Breeding in Aquaculture: An Introduction*,
Reviews: Methods and Technologies in Fish Biology and Fisheries 10,
DOI 10.1007/978-90-481-2773-3_8,

Table 8.1 Mass spawning of sea bream comprising of 40 female and 20 male parents. A sample of 1343 offspring were assigned to their parents by DNA analysis. (Rye pers. comm.)

Sire ID	No. offspring	Dam ID	No. offspring
63	391	58	266
40	285	38	162
24	182	36	140
11	176	34	138
		59	138
Proportion of offspring assigned to these sires	80%	Proportion of offspring assigned to these dams	63%

Spain by AFGC (Rye pers. comm.). In total, 40 females and 20 males were spawned in ponds. A batch of eggs was sampled from which 80% of the progeny were uniquely assigned to their parents. Representation of individual breeders showed that 80% of the progeny were assigned to four males and 63% to five females by DNA analysis. Representation of individual breeders are shown in Table 8.1.

Such unequal contribution from broodstock is also common in shellfish such as abalone, where even after efforts to normalize sperm content, certain sires were found to father the majority of offspring in one study (Selvamani et al. 2001).

By using a large number of broodstock, inbreeding problems can to a certain degree be avoided. However, the relative contribution of each breeder will be unknown and the effective population size will be reduced accordingly. For dissemination purposes, mass spawning is frequently used to produce large numbers of eggs and alevins.

This problems resulting from unequal parental contributions in mass spawning were discussed by Bentsen and Olesen (2002), who concluded that the number of progeny tested should be restricted and standardised to not less than 30–50 per broodstock pair, implying that some form of controlled mating is required.

8.3 Single Pair Mating

Production of full-sibs through the fertilisation of eggs from one female with milt from one male is the simplest mating design involving controlled mating. The number of families produced will be the same as number of male and female parents used (Table 8.2). If it is difficult to obtain sperm and eggs from broodstock by artificial stripping, single pair mating can be achieved through keeping pairs of breeders in separate tanks, ponds or hapas.

Bentsen and Olesen (2002) discussed the number of mating pairs needed to keep the inbreeding at an acceptable level, while at the same time obtaining a large genetic gains. As Fig. 8.1 shows, at least 50 pairs should be selected and tested in each generation, with no less than 30–50 progeny per pair. This design resulted in a selection

Table 8.2 Single pair mating

	Female		
Male	1	2	3
1	x		
2		x	
3			x

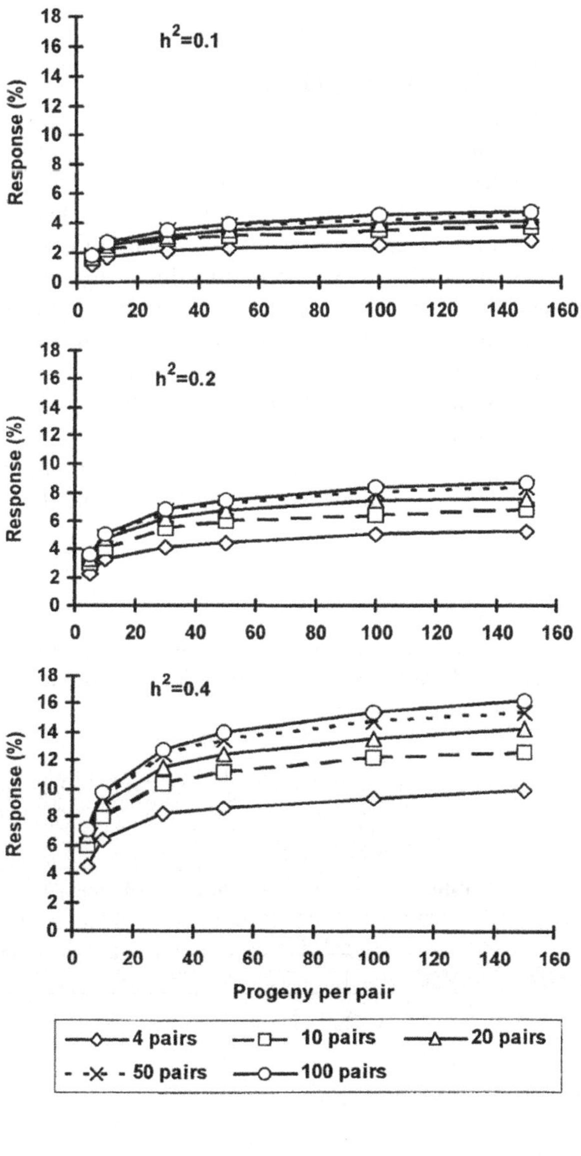

Fig. 8.1 Mean response to selection per generation (% of base population mean) over 15 generations of selection for different combinations of heritabilities (h^2), numbers of broodstock pairs selected, and number of progeny tested per pair, computed from stochastic simulations of 20 replicates for each combination. Reproduced from Bentsen and Olesen (2002) by permission of Elsevier

response of 5–13% of the base population mean, depending on the heritability of the trait. Increasing the number of broodstock to 100 pairs per generation increased genetic gain further. Reducing the number of broodstock pairs below 50 led to an increased rate of inbreeding of 6–8% per generation. This loss of genetic variation as the result of inbreeding was also found to reduce the response to selection by more than one third.

One of the weaknesses of the single pair mating design is that it is not possible to estimate how much of the genetic variance is additive and how much is of non-additive nature. Therefore this mating design is not optimal for use in a selective breeding program, but can still be effective for crossbreeding schemes.

8.4 Nested Mating Design

For aquatic species, a nested mating design (Table 8.3) is often used. In the nested mating design depicted in Table 8.3, male number 1 fertilises eggs from females 1 and 2, male number 2 fertilises eggs from females 3, 4 and 5 and male 3 fertilises eggs from females 6, 7 and 8. This results in eight full-sib families, the same as number of females used, representing three half-sib families, the same as the number of males used. In most species where artificial fertilisation is practiced, males could have been nested within females in the same manner. For the estimation of genetic variance components, it is preferable to have a mating design where females are nested within males, in order to minimise the possible confounding between additive genetic and maternal effects.

According to Gjerde (2005b), the estimation of genetic components can be performed using the following formula for sires and dams:

$$y_{fijk} = fixed_f + s_i + d_{ij} + e_{fijk} \tag{8.1}$$

where y_{fijk} is the value of a trait y, $fixed_f$ are effects like age, s_i is effect of sire, d_{ij} is effect of dam nested within sire and e_{fijk} is random error effect.

The genetic components are:

$$\sigma_s^2 = 0.25\sigma_A^2;\ \sigma_d^2 = 0.25\sigma_A^2 + 0.25\sigma_D^2 + \sigma_M^2 + \sigma_C^2 \tag{8.2}$$

Table 8.3 Nested mating design where females are nested within males

	Female							
Male	1	2	3	4	5	6	7	8
1	x	x						
2			x	x	x			
3						x	x	x

where the sire component (σ^2_s) represents ¼ of the additive genetic variance (σ^2_A), the dam component (σ^2_d) represents ¼ of the additive genetic variance (σ^2_A) together with ¼ of the dominance component (σ^2_D), possible maternal effect (σ^2_M), and common environment effect (σ^2_C).

The maternal effect (σ^2_M) represents possible effects transmitted from dam through her eggs, but this effect is usually considered to be insignificant in fish and shellfish. The common environmental effect or tank effect (σ^2_C) appears when each full-sib familiy have been reared in separate units for a period before the animals are physically tagged and subsequently pooled. Common environmental effects have been reported to account for around 4.5% of total variation in smolt weight in Atlantic salmon (Refstie and Steine 1978).

8.5 Factorial Mating

A factorial mating design is illustrated in Table 8.4. Eggs from one female are divided into portions and each portion is fertilised with milt from different males. In this case, 29 full-sib families are produced from nine males and nine females. The factorial design produces both paternal and maternal half-sibs. A downside to the factorial mating design is that, for a given number of rearing tanks/hapas available, a lower number of broodstock can be tested than in a nested design.

The formula for the animal model according to Gjerde (2005b) is:

$$y_{fijk} = fixed_f + a_i + fs_j + e_{fijk} \tag{8.3}$$

where y_{fijk} is the value of trait y, $fixed_f$ are fixed effects attached to the trait, a_i is the additive genetic effect, fs_j is the effect of full-sibs caused by dominance, maternal and common environmental effects and e_{fijk} is the random error effect.

The genetic components are:

$$\sigma^2_a = \sigma^2_A;\ \sigma^2_{fs} = 0.25\sigma^2_D + \sigma^2_M + \sigma^2_C \tag{8.4}$$

Table 8.4 Factorial mating design

	Female								
Male	1	2	3	4	5	6	7	8	9
1	x	x							
2	x	x							
3			x	x	x				
4			x	x	x				
5			x	x	x				
6						x	x	x	x
7						x	x	x	x
8						x	x	x	x
9						x	x	x	x

where σ^2_A is the additive genetic effect and σ^2_{fs} contains dominance, maternal and common environmental effects.

8.6 Connectedness

The genetic gain obtained in a breeding program depends to a large extent on the quality of the data used for estimation of breeding values. If the genetic groups are tested in different environments there must be some genetic ties between the environments in order to connect the data mass. Data can be disconnected for several reasons:

- Families are reared in different test units
- The breeding population consists of groups of families (cohorts) tested at different times of the year
- Families are exchanged between breeding populations.

Connectedness can be ensured by creating direct ties, i.e. having some breeders with progeny in more than one environment, or across year classes (Fig. 8.2).

Direct genetic ties may be required for unbiased estimation of genetic gain. Multiple spawners can be used for repeated mating of breeders that have been used earlier, and other special control populations are also frequently used. Methods for measuring genetic changes are discussed in Chapter 11.

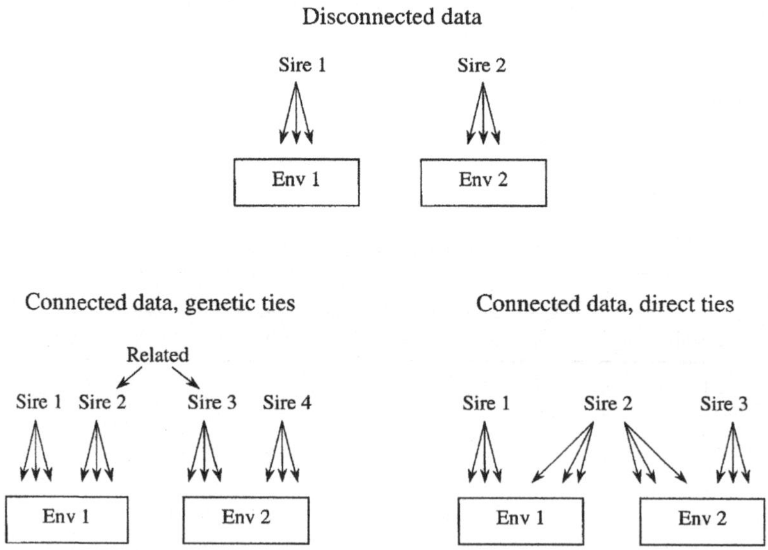

Fig. 8.2 Illustration of data that are disconnected and data that are connected through genetic ties or direct ties. Reproduced from Gjerde (2005b) by permission of Springer

8.7 Conclusion

No single mating design is optimal for all situations. However, regardless of the mating design chosen, inbreeding poses a serious problem in any breeding program and therefore must be controlled. This is typically achieved by applying controlled mating and separate rearing of families until the offspring can be physically tagged. This requires a large number of rearing units in the hatchery. Alternatively, DNA markers can be used to determine parentage retrospectively among offspring produced by mass spawning. This approach eliminates common environmental effects (σ^2_C) discussed earlier, as the offspring of different families can be communally reared throughout the production cycle.

For a given testing capacity, the advantage of a single pair mating design is that the number of tested broodstock is twice the number of families, and hence produces a low rate of inbreeding (Gjerde 2005b). The weakness of the single pair mating design, however, is that it is not possible to separate additive and non-additive genetic variance, and therefore this method is not preferred in most aquaculture situations.

Compared to the single pair mating design, a factorial mating design leads to faster accumulation of inbreeding since fewer breeders are used. Their advantage is that both additive and non-additive genetic effects can be accurately estimated.

The nested mating design has the advantage of allowing more broodstock to be tested than the factorial design, and hence the accumulation of inbreeding is lower. Like the factorial mating design, the nested design allows for accurate estimation of additive genetic effects, but not non-additive genetic effects.

Considering the advantages and disadvantages of these three mating designs for a given testing capacity, a nested mating design offers the most benefits with the fewest disadvantages, and therefore tends to be the preferred option.

Chapter 9
Estimation of Breeding Values

9.1 Introduction

To efficiently apply selection, a measure of the gene alleles that are passed on from parents to progeny is required, since the true genotype of an individual cannot be measured. This measure is known as the breeding value. In genetic terms, the breeding value of an individual is equivalent to the mean value of its progeny, and is considered to be unique to the population in which it will be mated (Falconer and Mackey, 1996).

The purpose of a breeding program is to increase productivity of a population, which in real terms means the moving of the averages of economically important traits in a desired direction. This is achieved through the ranking of potential breeders, culling poor performers and selecting the highest-ranking animals as broodstock for subsequent generations. For maximising genetic gain it is vital that potential breeders can be ranked as accurately as possible, or in other words, their breeding values are predicted as accurately as possible. The accuracy of a breeding value is defined as the correlation between the true (the genetic constitution of an animal) and the predicted breeding value.

To be able to accurately predict the breeding value of animal, it is therefore necessary to have records of all traits in the breeding goal. Production records may be obtained from the individuals themselves, full- and half-sibs, progeny and parents. Records from relatives can be used because they share common alleles with the breeding candidate, and the increased number of shared alleles between close relatives means that close relatives are better genetic predictors than more remote relatives. Records of full-sibs are thus more valuable than records of half-sibs (unless common environmental effects for full-sibs are large). In addition, estimates of genetic and phenotypic variances, heritabilities and phenotypic and genetic correlations between the traits are required in order to construct reliable selection indexes. Additional pedigree information is needed to control the accumulation of inbreeding in the population under selection.

The breeding value of an individual cannot be measured directly, or with 100% accuracy unless information from an indefinite number of progeny is available (Table 7.1). However, in practice this is of course impossible to obtain. The true

T. Gjedrem, M. Baranski, *Selective Breeding in Aquaculture: An Introduction*,
Reviews: Methods and Technologies in Fish Biology and Fisheries 10,
DOI 10.1007/978-90-481-2773-3_9,

breeding value will therefore be unknown, and can only be estimated from the product of the individual gene effects manifested as the phenotypic value of the trait.

9.2 Breeding Value of Individual Animals

Selection of broodstock should be based on breeding values. According to classical selection index methodology, estimation of breeding values should be based on records pre-adjusted for systematic environmental effects such as differences in age, sex, and environment among farms and ponds/cages. Such adjustments are discussed in Section 5.6. By eliminating these environmental effects, the accuracy of breeding values will be increased. However, data need not be adjusted beforehand if solutions to Henderson's mixed model equations are used for estimation of BLUP breeding values (see Section 9.7), since the MME methodology allows for simultaneous inclusion of fixed effects.

The breeding value, A_i, of an animal based on one trait (X) can be estimated as:

$$A_i = h^2(X_i - \bar{X}_i) \quad (9.1)$$

where h^2 is the heritability of the trait, X_i is record of the ith animal and $\bar{X}_i$ is the population average. Thus, when predicting breeding value based on one trait, the weighing factor is the heritability of that trait. The variation in breeding values among animals depends largely on the magnitude of the heritability and standard deviation of the trait.

According to Van Vleck et al. (1987), the accuracy of estimated breeding values is the square root of the heritability (h).

9.3 Breeding Value of Full-Sib Families

The average genetic relationship, r_G, between full-sib family members is:

$$r_G = 0.5 \quad (9.2)$$

Family selection is based on estimates of the genetic value of a randomly picked member of a family from the phenotypic average of all its members. The greater the number of animals contributing to the family average, the closer the average family phenotype approaches the true genetic merit of the family, or its additive genotype. This is because the environmental components, after the systematic effects have been adjusted for, are generally random in nature and will approach zero as the family number increases.

In breeding programs for aquatic animals, families are frequently produced and reared communally after tagging. It is therefore of interest to rank the families according to their estimated average breeding values. Family information can be used to substantially increase breeding values of individuals. The use of family records is also of great importance when selecting for traits that may not be recorded on the breeding candidate like sex limited traits, carcass quality traits and traits that only can be quantified in frequencies like mortality, disease resistance, and seasonal sexual maturation.

Breeding values (A_i) of full-sib families (FS) using averages of one record per member can be estimated as:

$$A_j = nh^2(X_{jn} - \bar{X}_{jn})/(2 + (n - 1)(h^2 + 2c^2_{FS})) \quad (9.3)$$

where X_{jn} is the record of trait X of ith individual and n number of animals in the family, $c^2{}_{FS}$ is maternal effect.

9.4 Breeding Value of Half-Sib Families

The genetic relationship, (r_G), between half-sibs is:

$$r_G = 0.25 \quad (9.4)$$

The breeding value (A_i) of a half-sib family (HS) with average records of n family members is:

$$A_j = nh^2(X_{jn} - \bar{X}_{jn})/(4 + (n - 1)(h^2 + 4c^2_{HS})) \quad (9.5)$$

where X_{jn} is the record of trait X of ith individual and n number of animals in the family, $c^2{}_{HS}$ is maternal effect.

9.5 Breeding Values for Multiple Traits Using a Selection Index

Selection indexes were first applied in plant selection programs by Smith (1936) and later developed for terrestrial livestock by Hazel (1943). Numerous studies have subsequently extended the theoretical foundation of the selection index approach and its application. Hazel and Lush (1942) concluded that selection for all traits simultaneously using an aggregate selection index is superior to other methods, and this is particularly true in a complex breeding program selecting for several traits and with information of individual records together with information from full-sibs, half-sibs and ancestors. Falconer and Mackay (1996) state that a selection index is the best linear prediction of an individual's breeding value and takes the form of a multiple regression of breeding values on all the sources of information.

Construction of a selection index requires knowledge of phenotypic and genetic parameters for all traits involved. The central parameters are the average, standard deviation and heritability of each trait, phenotypic and genetic correlations between traits, and the relative economic weight to be assigned to the individual traits included in the breeding goal. For further optimisation of multi-trait selection programs in fish and shellfish species, more accurate and reliable genetic and economic values of important traits are needed for most species. Some examples of genetic parameters are given in Tables 4.1 and 5.1.

A selection index may be expressed as:

$$I = b_1(X_1 \bar{X}_1) + b_2(X_2 - \bar{X}_2) + +b_n(X_n - \bar{X}_n) \quad (9.6)$$

where b_i is the standard partial coefficient of regression, n is number of traits and the different Xs are expressed as deviations from the population averages ($\bar{X}$). The b_i values are weighted according to the relative economic importance of each trait. By weighting each genotype by its relative economic value, they together add to what is known as the aggregate genetic value (H). Chapman (1962) defined relative economic values as: 'A unit change in the additive value of the genotype as far as characteristic (1) is concerned would have a value in terms of profitability (a_1), where a unit change in the additive genetic value for characteristic (2) would have an effect on profitability of (a_2)'. It is therefore important to estimate the relative economic value (a_i) precisely.

It has been shown that all traits of economic importance should be included in the aggregate genotype (Gjedrem, 1972). Even economically important traits that are not recorded should be included in the aggregate genotype. An example of such a trait is feed conversion efficiency which is difficult and expensive to record in fish and shellfish. Feed conversion efficiency is, as shown in Section 5.4.3 highly correlated with growth rate. When including feed conversion efficiency and other economically important traits in the aggregated genotype, the trait will receive the correct weighting and the genetic gain will be increased relative to excluding them from the aggregate genotype. This is particularly true when the trait in question is highly correlated with one or more traits in the breeding goal (Gjedrem, 1972).

Gjedrem (1967) showed that there is potential to achieve gain by including correlated traits or genetic markers in the selection index. These correlated traits may have no economic value. The maximum gain will be achieved if the economic trait has a low heritability and the increased genetic gain is highest when the genetic and environmental correlations between the traits are high and negative. If the economic trait has a high heritability, the greatest potential for gain occurs when the correlated trait has low heritability.

9.6 Scaling of Selection Indexes

For reporting purposes it is preferable to transform the absolute index values to a fixed mean and standard deviation. For example, index values transformed to have an average of 100 and a standard deviation of around 10 would yield indexes ranging from around 70–130. The standardised index values have the advantage that they are unaffected by changes caused by shifts in the relative economic values and population means of the traits.

9.7 Best Linear Unbiased Prediction (BLUP)

When all available information about an individual as well as information from full- and half-sibs, progeny and ancestors are used to estimate breeding values, the calculations become very complicated (described in detail by Gjerde (2005a)). A more powerful procedure is the Best Linear Unbiased Prediction (BLUP) method developed by Henderson (1975). In animal and plant breeding, this method was first implemented in the early 1970 s and is now considered to be the best method for the estimation of breeding values. This method is now frequently used for aquatic species, and estimates systematic environmental effects and breeding values simultaneously. A BLUP model that accounts for all the genetic relationships among animals whose breeding values will be estimated is known as an animal model.

Chapter 10
Genotype–Environment Interaction

10.1 Introduction

Environmental conditions for aquaculture production vary considerably from country to country and from one climatic zone to another. This means that stocks that are better adapted to local environmental conditions tend to perform better, resulting in a wide variety of different species being farmed across the world. Water temperature is one of the key environmental parameters that dictates the selection of species to be farmed, in the tropics species tolerating high temperatures are farmed while cold water species are predominantly farmed in temperate climates. Some environments are particularly prone to large variations in conditions, in such situations hardy species with broad tolerance limits are favoured.

The difference in performance of a species across environments is a very important consideration in selective breeding programs. Substantial genotype-environment interactions (G×E) from a breeding perspective implies that genetic groups or individuals rank differently in different environments (Gjedrem 2005). Such re-ranking of individuals and families will reduce the genetic gains obtained in breeding programs serving larger production regions. Some species are particularly sensitive to environmental changes, and even more tolerant species can be significantly affected once environmental conditions reach a certain threshold. Therefore, an important selection criteria for choice of species as well as breeding goal is the animals' robustness and ability to tolerate variation in environmental conditions.

In terrestrial livestock species, it has been found that narrow breeding goals such as those focusing solely on milk yield in dairy cattle and growth rate in broilers have developed animals that are highly sensitive to environmental conditions. Selecting broadly, i.e. including several traits in the breeding goal is one way to increase the robustness of animals, and therefore is a recommended strategy for aquatic species.

10.2 Estimates of Genotype–Environment Interactions

To study G×E effects, different genotypes must be tested under different environmental conditions. For Atlantic salmon and rainbow trout, extensive studies of

T. Gjedrem, M. Baranski, *Selective Breeding in Aquaculture: An Introduction*,
Reviews: Methods and Technologies in Fish Biology and Fisheries 10,
DOI 10.1007/978-90-481-2773-3_10, © Springer Science+Business Media B.V. 2009

G×E interactions were performed by AKVAFORSK in the 1970s. For three years, 35 strains of Atlantic salmon were reared at five locations spanning a range of approximately five degrees of latitude. Results demonstrated considerable variation in growth rate between locations, reflecting differences in water temperature and light conditions. As shown in Table 10.1, however, the farm by strain interaction effect accounted for only 1.4–3.7% of the total variation in growth (Gunnes and Gjedrem 1978). A similar study was conducted for rainbow trout, with 94 sire progeny groups tested at up to five locations over a four year period with a latitude difference between the southern and northern farms of seven degrees. The variation in growth rate between farms was considerable and the sire by farm interaction effect for growth ranged from 1.2 to 5.5% of the total variance for body weight (Gunnes and Gjedrem 1981) (Table 10.1).

These results showed that G×E effects tend to account for only a small portion of total variation in growth rate even if the animals were subject to large variation in seawater temperature and day length. These findings lead to the conclusion that it is not necessary to develop specific strains of Atlantic salmon and rainbow trout for farming in the northern and southern regions, respectively.

An alternative approach for assessing the effect of G×E interactions is estimate the genetic correlation between a trait recorded in different environments, considering them as separate traits. Table 10.2 presents such estimates based on analyses of harvest weights recorded in different environments for Atlantic salmon and rainbow trout. Most of the genetic correlations are high (> 0.70), indicating negligible or marginal effects of G×E interaction. However, an unexplained low genetic correlation was observed for Atlantic salmon in year class 1982. A low genetic correlation was also observed in rainbow trout reared in fresh and brackish water ($r_G = 0.58$) which may reflect a real G×E interaction where families are ranked very differently in these two environments that differ in salinity. However, this explanation is somewhat contradicted by the observation that full saline (A+T) and zero salinity (B) environments yielded a genetic correlation of 0.86.

Table 10.1 Genotype–environment interaction as a percentage of the total phenotypic variation for Atlantic salmon and rainbow trout

Species	Source of variation	Year class			
		1972	1973	1974	1975
Atlantic salmon	Between strains	7.0	6.7	8.6	
	Interaction strain-farm	3.7	1.5	1.4	
Rainbow trout	Between sires	22.6	4.7	7.3	12.2
	Interaction strain-farm	2.3	5.5	3.8	1.2

Reproduced respectively from Gunnes and Gjedrem (1978) and Gunnes and Gjedrem (1981) by permission of Elsevier

Table 10.2 Genetic correlations between harvest weights in different farming environments

Species	Environment[1]	Year class	Genetic correlation	Reference
Atlantic salmon	A – T	1978	0.89	McKay and Gjerde (1986)
	A – T	1980	0.79	Standal and Gjerde (1987)
	A – T	1981	0.84	Standal and Gjerde (1987)
	A – T	1982	0.52	Standal and Gjerde (1987)
Rainbow trout	A – T1	1984	0.82	Sylven et al. (1991)
	A – T2	1984	0.79	Sylven et al. (1991)
	T1 – T2	1984	0.86	Sylven et al. (1991)
	A+T – B	1984	0.86	Sylven et al. (1991)
	A + T – C	1984	0.72	Sylven et al. (1991)
	B – C	1984	0.58	Sylven et al. (1991)

[1]A = AKVAFORSK's research station at Averøy, salinity < 30‰; T = AKVAFORSK's test stations (private farms), salinity < 30‰; B = Fish farm in Sweden, salinity zero; C = Fish farm in Sweden, salinity 4–8 ‰

For common carp, G×E interaction effects have been extensively studied in Israel. In general, considerable G×E interactions were observed that resulted in substantial re-ranking of genetic groups. Moav et al. (1975) studied growth rate in 12 different genetic groups of common carp, reared in five different pond environments. The ranking difference was particular large for the extreme genotypes between the extreme environments. Wohlfarth et al. (1983) compared the ranking of three genotypes of common carp (Chinese and European strains and their hybrid) in five different environments and found considerable G×E interaction.

Reddy et al. (2002) studied G×E interaction in rohu carp in India by rearing different strains in monoculture and polyculture (Fig. 10.1). In 1993, two river strains were tested and the G×E interaction effect for harvest body weight accounted for less than 1% of the total variation. In 1994, six river strains were tested under these two environmental conditions. As can be seen in Fig. 10.1 the ranking was very similar in monoculture and polyculture systems for all strains and the G×E interaction accounted for only a minor portion of the total variation. Similar results were obtained for survival between tagging and harvest.

A similar, extensive investigation of G×E interactions for growth rate of tilapia was undertaken in the GIFT project in the Philippines. Seven strains of Nile tilapia were reared in 11 highly different environments, with mean recorded body weights varying from 121 g in the best environment to 9 g in the poorest. In Fig. 10.2, the environments are ranked from left to right for increasing growth rate, and strains are ranked within each environment as a deviation from the strain average. The results showed that the overall ranking of strains across environments was highly consistent. In environments with low growth rate, there were some differences in

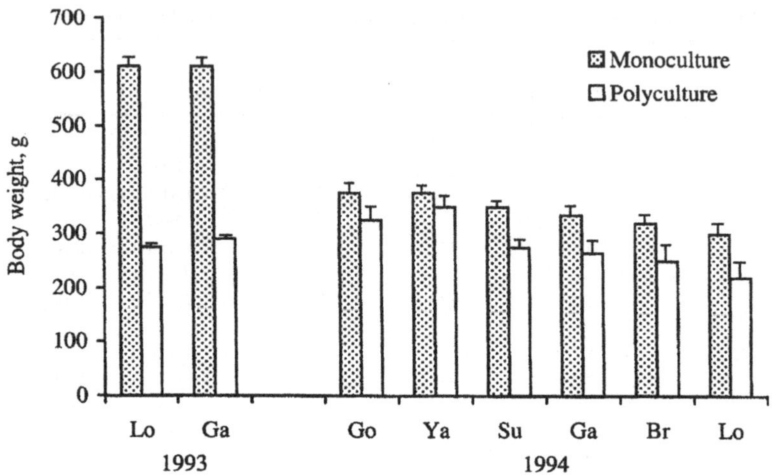

Fig. 10.1 Harvest body weight for rohu carp in monoculture and polyculture. Ranking based on body weight in monoculture within each year class. Reproduced from Reddy et al. (2002) by permission of Elsevier

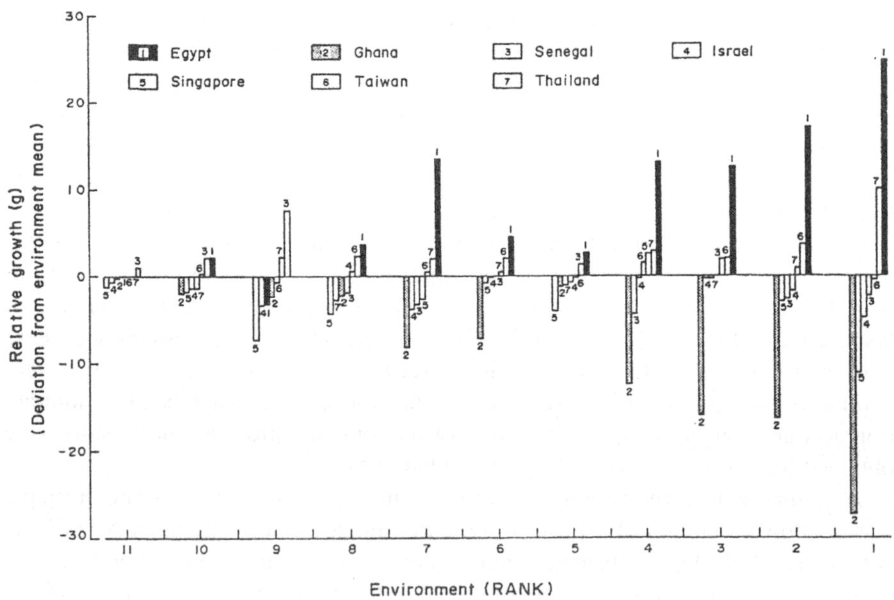

Fig. 10.2 Relative growth performance of seven strains of Nile tilapia in different test environments. Relative growth performance was calculated as the deviation of the final body weight mean of each strain from the mean of the particular environment. Reproduced from Eknath et al. (1993) by permission of Elsevier

rank order for some of the strains. For the total dataset, G×E interaction effects accounted for only 0.3% of the total variation in body weight in this study. It is interesting to note that the differences between strains within each environment increased markedly in the higher growth environments.

The tilapia species *Oreochromis shiranus* was used to study G×E interaction for growth rate in Malawi (Maluwa et al. 2006). In total, 166 full-sib families were tested in three farms located at different altitudes. All genetic correlations between body weight in different test environments were not significantly different from unity, varying from $r_G = 0.63 - 0.95$, suggesting no G×E interaction. The authors argued that more reliable genetic correlations between harvest body weights in different test environments should be obtained using replicated ponds at each test environment before firm conclusions are drawn on possible G×E interaction effects in this species.

Dupond-Nivet et al. (2008) studied G×E interactions for growth rate in European sea bass, based on data from 253 full-sib groups reared at four farm locations in France. The genetic correlations between harvest weight recorded at the different farms were all very high (≥ 0.84), except for one estimate of 0.70.

Swan et al. (2007), studied G×E interactions for growth rate in Pacific oysters in South Australia where full-sib families were reared in five environments. Apart from one environment, the genetic correlations between environments averaged $r_G = 0.90 - 0.91$, indicating that G×E interactions in the form of re-ranking of families across environments were negligible. Using body weights measured at five time points between one and two years of age, estimates of genetic correlations between the four final weights were all above $r_G = 0.96$ and only slightly lower between the first weight and at older ages. This led to the conclusion that G×E interactions would not be detrimental to the breeding program for the tested population. However, these results were in sharp contrast to the findings of Langdon et al. (2000), where strong G×E interactions for growth rate were demonstrated for Pacific oyster on the west coast of the USA.

In shrimp, Fjalestad et al. (1997) studied genotype–farm interactions in Hawaii using 294 full-sib families tested at three to four farms. The family–farm interaction accounted for 0.1–8% of the total variation. Suarez et al. (1999) tested 52 full-sib families of *Penaeus vannamei* in two environments in Colombia. A strong genetic correlation ($r_G = 0.87$) between the ranking of the genetic groups in the two environments was found for growth rate, representing a low level of G×E interaction. A lack of significant G×E interaction has also been observed for growth rate in scallops in Mexico, (Ibarra et al. 1999) and in abalone in Iceland (Jonasson et al. 1999).

The ability of rainbow trout to efficiently utilise plant-based diets was investigated by Pierce et al. (2008). Fish fed fishmeal diet were around 8% larger than fish fed the plant-based diet, and a significant G×E interaction was found that accounted for 5% of the total variance. The genetic correlation between growth rate for the two diets was $r_G = 0.73$. It was concluded that there is substantial genetic variation for the utilisation of plant-based diets containing soybean meal and oil in rainbow trout. This is in agreement with the findings at AKVAFORSK of Austreng and

Refstie (1979) where a near significant interaction between protein levels in diets and families of rainbow trout was found. On the contrary, Edwards et al (1977) did not find a significant interaction between carbohydrate content of diets and families of rainbow trout.

10.3 Conclusion

Generally, from the published literature to date, it appears that G×E interactions account for only a small part of the total variance in growth rate and survival. This has been found to be the case even when strains with large genetic differences have been tested under very different environmental conditions, as was the case for tilapia project in the Philippines. However, the examples given show that the magnitude of the interaction estimates vary greatly and that it is difficult to make general conclusions regarding this effect. It is clear that in some situations there are considerable G×E interactions that must be accounted for when planning breeding programs. Therefore, it can be strongly recommended to evaluate the magnitude of at least the most obvious potential G×E interactions foreseeable in a breeding program.

If G×E interactions account for a considerable portion of the total variation, separate breeding programs should be developed for each environment where major re-ranking of genetic groups is seen to occur. If a breeding company decides to start an additional program to meet such interactions, the testing capacity should be doubled in order to obtain satisfactory genetic gain. However, this could potentially double the running costs of a breeding program.

Another lesson to be learned from the examples given of G×E interactions, is that families should be tested under varying practical farming conditions common to the population in question. This will ensure that breeding candidates can be ranked according to their average genetic merit across the range of relevant production environments, and possible interaction effects can be taken into account. This approach favours development of robust genotypes that tolerate variable environmental conditions. Simply testing the families under standardised environmental conditions at a breeding station may not account for possible G×E interactions in the field where grow-out takes place, and may significantly reduce the genetic gains expressed under commercial farming conditions

As mentioned earlier, a broad breeding goal may be a means to develop robust animals with high tolerance to variable environmental conditions. However, it should be kept in mind that increasing the number of traits in the breeding goal will reduce the amount of genetic gain that can be obtained from each individual trait. To find the best balance between these factors is a challenging task and requires expertise and experience by the breeders.

Chapter 11
Measuring Response to Selection

11.1 Introduction

In Section 4.9.4, the formula for predicting genetic response is given for individual selection for a single trait. This prediction relies on prior knowledge of the selection intensity (i), heritability (h^2) and standard deviation (σ_P) for the trait in question. The prediction describes what can be expected when selection is performed, and provides a good guideline when the parameters used are reliable. In practical terms, however, it is more valuable to know the magnitude of the realised genetic response that can be obtained in the breeding program. This requires a different, and much more difficult, approach. The major problem associated with calculation of realised genetic gain is that changes in phenotypes measured over time are the sum of genetic and environmental changes, and possible G$\times$E interactions.

There are several important reasons for why a breeding program should implement methods for monitoring realised genetic gain:

- To document the magnitude of genetic gain for each trait in the breeding goal
- To assess whether the predicted genetic gain is actually being reached
- To identify the main factors responsible if this predicted gain is not reached.

As more traits are included in the breeding goal, more resources are needed to obtain reliable estimates of genetic gain for the individual target traits. Furthermore, as the number of traits involved increases, it becomes more complicated to obtain good estimates of the genetic gain for each trait.

To obtain estimates of realised genetic gain, some of the available testing capacity must be used. When the testing capacity is limited, a decision will have to be made to what extent the resources should be used for testing new families and to what extent they should be used for estimating genetic gain. This will always be a central question when planning a breeding program. To date, there has been relatively little focus on estimating realised genetic gain in breeding programs for aquatic species. The lack of documented genetic gains and the economic benefit from using improved stocks may be one of the reasons for the slow rate of implementation of breeding programs in aquaculture industries.

T. Gjedrem, M. Baranski, *Selective Breeding in Aquaculture: An Introduction*,
Reviews: Methods and Technologies in Fish Biology and Fisheries 10,
DOI 10.1007/978-90-481-2773-3_11,

There are several methods to measure response to selection and some of the most effective will be discussed in the following sections. More extensive discussions are found in Gall et al. (1993) and Rye and Gjedrem (2005).

11.2 Control Population

Traditionally, a randomly reproduced control line has been used to estimate genetic change. This relies on the assumption that any changes in environmental conditions will equally affect the control line as well as the selected line. The control line should be derived from the same base population as the selected line, and the two lines must be tested under the same environmental conditions, implying communal rearing in the same pond or in the same cage.

In practice, this means that samples from all families should be reared together with the control line. If the control line is to be used for several generations, it is important to use many pairs of broodstock in each generation to avoid the accumulation of inbreeding within the line.

The response to selection (ΔG) is estimated as the difference in mean performance between the selected ($P_{selected}$) and control line ($P_{control}$):

$$\Delta G = P_{selected} - P_{control} \tag{11.1}$$

The premise for this equation is that the environmental conditions for the two lines has been equal during the whole lifespan including the grow-out period which in practice means that the two lines have been reared communally.

As selection proceeds over generations, the difference between the lines will become larger and G×E interactions may become significant. This will bias the estimate of the response to selection. Usually the effective population size (Ne) of the control line will be small and thus subject to random drift which may cause random changes in allele frequencies. This will limit the validity of a comparison between lines after few generations, as discussed by Gall et al. (1993). This implies that the use of an unselected control line may be a good method for estimating genetic change in the first two to three generations of selection but other methods are superior in a long-term breeding program.

11.3 Average Breeders

The use of average breeders is a method commonly applied for both fish and shellfish, and is illustrated in Fig. 11.1. A number of breeders with average estimated breeding values are used to produce a pool group of offspring that serve as a control group for the following generation. The reasoning behind this approach is that the offspring produced from the average breeders are assumed to represent the average genetic level in the parent generation. The response to selection is based on the dif-

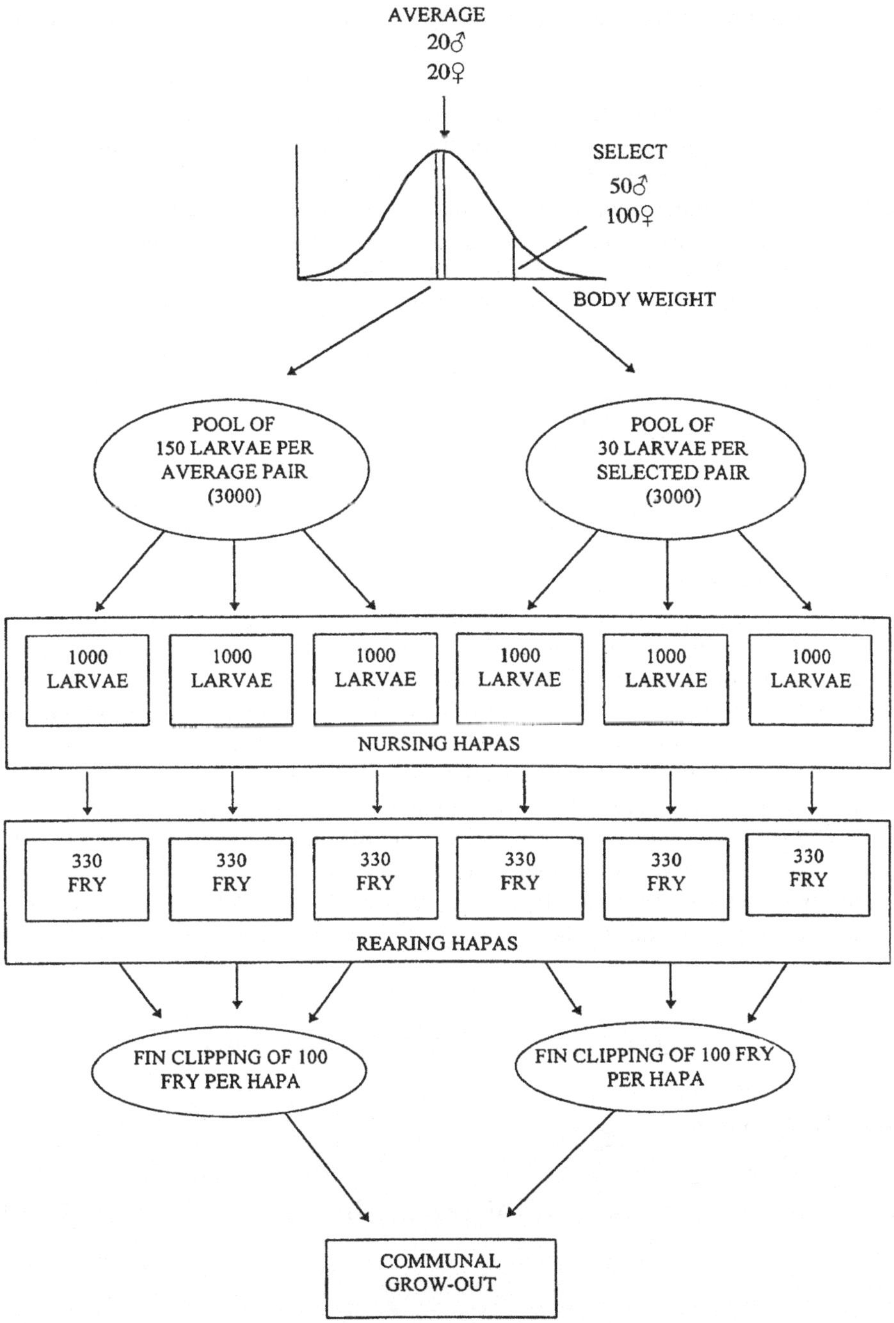

Fig. 11.1 Control of genetic gain. Comparison of progeny of selected breeders with progeny of average breeders. Reproduced from Gjedrem et al. (1997) by permission of WorldFish Center

ference between the average performance of offspring between the selected and the average breeders.

Figure 11.1 reflects the scenario of mass selection for growth rate in tilapia but can also be applied to multi-trait selection. If there are many traits in the breeding goal, the breeding value must be estimated by means of a selection index. The method may also be used for other species where tanks or ponds may replace hapas and is not resourse demanding because the pool for the selected group can be sampled from each family under test. The number of pairs of average breeders used to produce the pool of the control group must be sufficiently large to avoid sampling variance that may bias the estimate of selection response. Expected genetic gain, ΔG is:

$$\Delta G = P_{\text{selected}} - P_{\text{aveage breeders}} \tag{11.2}$$

11.4 Repeated Matings

The comparison of contemporary animals belonging to different generations provides a means of estimating changes without using traditional control populations. Repeated mating in successive generations can be implemented for multiple spawners and is possible for many fish and shellfish species. For species spawning only once, cryopreservation of sperm may allow repeated mating.

The approach compares offspring of males belonging to different generations when the male was mated to a random sample of females each year. Let the offspring from the first year of mating be P_1 and the second year P_2, reared under the same environmental conditions, E_1 and E_2, respectively. The difference between the groups will represent half of the genetic gain since the sire represents only half of the genetic contribution:

$$1/2\Delta G = P_2 - P_1 = (G_2 + E_2) - (G_1 + E_1) \tag{11.3}$$

when $E_1 = E_2$ then

$$1/2\Delta G = G_2 - G_1 \tag{11.4}$$

and

$$\Delta G = 2(G_2 - G_1) \tag{11.5}$$

11.5 Genetic Trend Analysis

Some statistical methods allow the separation of genetic and environmental effects. In 1975, Henderson developed the mixed model that allows estimates of genetic trends. The accuracy of the estimates depends on the genetic connectedness between individuals across environments and year. Mixed models use all relationships between available animals to estimate breeding values, and if the data comprises information from more than one generation, genetic trends from selection can be estimated. Gall et al. (1993) concluded that the mixed model approach could become a powerful tool to analyse fish breeding data in the future.

11.6 Conclusion

Given the biological nature of animal breeding, fluctuations between years are to be expected, particularly for small breeding units. Therefore, accurate and reliable estimates of genetic gain can only be obtained after several generations of selection.

The estimation of realised genetic gain is an important component of all breeding programs supplying eggs/fry to the aquaculture industry, since the magnitude of genetic gain provides a key measure of the efficiency of the breeding work. Such estimates can also be valuable in identifying necessary changes to be implemented if predicted gains are not achieved.

The optimal choice of method to measure realised genetic gain may differ between species as well as between stages of a breeding program. For the first two to three generations of selection, control populations may be an appropriate method while for long-term breeding programs, average breeders or repeated mating approaches tend to be more appropriate. Genetic trend analysis will be a good alternative when complete pedigree information is available for several generations of selection and genetic ties are continuously produced.

Chapter 12
Structure of Breeding Programs

12.1 Introduction

Running a breeding program with its many inherent elements is not a simple task, and requires a great deal of organisation and structure to be successful. The leadership within the program needs to be well-qualified in the areas of administration and economics, together with at least a basic knowledge of breeding theory. Those involved in running the technical parts of the breeding work need an extensive knowledge of quantitative genetics and selective breeding theory. Although any breeding program must be started on basis of a comprehensive and detailed plan based on present knowledge, continuous improvement reflecting experiences and new research is needed as the program develops.

Compared to terrestrial livestock species, the application of selective breeding in aquatic species still remain in its infancy, and for many species basic knowledge needed to plan and develop efficient breeding programs is lacking. As a result, significant research efforts continue to be needed with respect to traits of interest, genetic parameters, economic values, interactions and the many other important factors underlying successful implementation of a breeding program. Although breeding programs for some aquatic species, like salmon and tilapia, have proved to be extremely successful with documentation of impressive genetic gains, there is still an enormous untapped potential for genetic improvement in all the other species farmed across the world.

12.2 Breeding Programs Applying Individual Selection

Individual selection, or mass selection, can be applied for traits that can be measured on live breeding candidates. In most cases, this means that body weight (or a related growth trait) is the trait targeted by such selection. Individual selection is most effective for traits with medium to high heritability (e.g. growth rate), but is usually inefficient for binary traits like survival and sexual maturation.

Individual selection for growth is often the initial focus of breeding programs for small populations, however even large-scale programs may begin in this manner

T. Gjedrem, M. Baranski, *Selective Breeding in Aquaculture: An Introduction*, Reviews: Methods and Technologies in Fish Biology and Fisheries 10,
DOI 10.1007/978-90-481-2773-3_12, © Springer Science+Business Media B.V. 2009

in order to obtain practical experience with the operations involved. Activities may later be gradually extended to a larger scale and eventually include selection for several traits.

When a breeding program is initiated in a region where the industry is in its infancy, it is crucial to demonstrate the rate of progress and the economic advantage that can be achieved through selection as soon as possible. Applying individual selection for growth rate over two or three generations could be an effective method to rapidly demonstrate the value of the program. As shown in Chapter 3, growth rate typically responds very fast to selection and the positive results may soon be evident to the farmers.

A highly effective mass selection program for growth rate can be run without physical tagging of the animals. This approach saves the costs associated with rearing families separated until tagging, and avoids introduction of common environmental effects. If such a strategy is applied, however, a high number of broodstock pairs in each generation are required to avoid a rapid accumulation of inbreeding (Fig. 12.1). A random sample of eggs should therefore be incubated and hatched from each pair separately, and the number of progeny stocked for testing from each pair should be sufficient to ensure that an average of at least 30–50 progeny per pair complete the test.

A breeding program without identification of animals and no standardisation of numbers in each progeny group will usually lead to a rapid accumulation of inbreeding. If selection is performed for a highly heritable trait like body weight, the highest ranking animals tend to originate from relatively few families and consequently mating of close relatives will take place, as discussed in Chapter 8 (Fig. 12.2).

12.3 Advanced Breeding Programs

When there is sufficient scope for breeding to be undertaken on a larger scale in a given species, a combination of individual and family selection is preferred. This combined approach facilitates efficient selection for several traits. In this context, a full-scale breeding program may be implemented in a number of different ways. The most common design is illustrated in Fig. 12.3 and is discussed in the following sections mainly in the context of a fish breeding program.

The central part of such a program is the breeding nucleus where production and testing of families takes place, and selection of broodstock is carried out using physical tags for identification. The physical unit where the breeding work is carried out is called the breeding station.

The most comprehensive but expensive solution involves the construction of a dedicated hatchery unit with separate hatching trays for each family and separate units for rearing fingerlings of individual families until they reach tagging size (typically minimum 5–10 g). As the number of tested families in each generation should

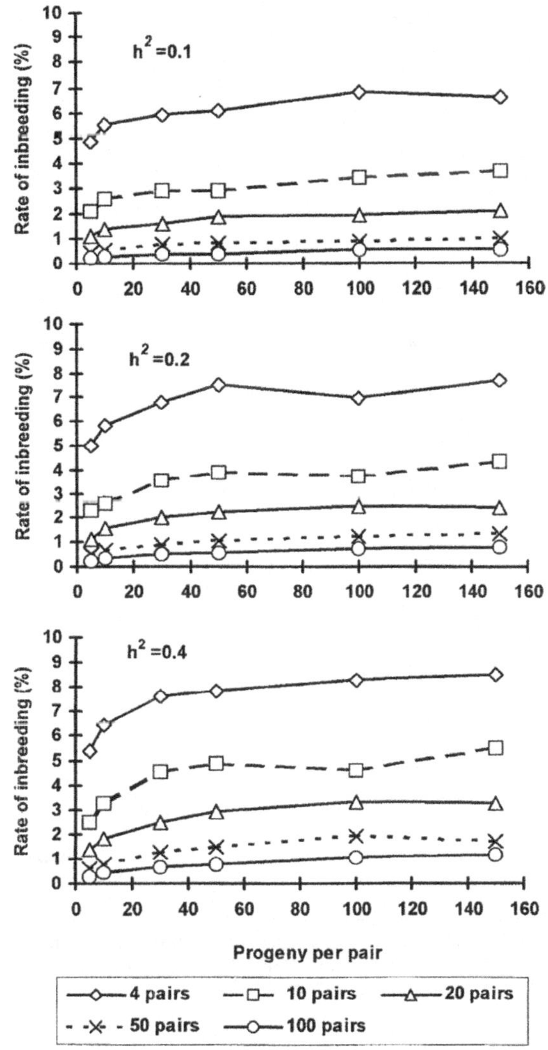

Fig. 12.1 Mean rate of inbreeding per generation (%) over 15 generations of mass selection for different combinations of heritabilities (h^2), numbers of broodstock pairs selected and numbers of progeny tested per pair, computed from stochastic simulations of 20 replicates for each combination. Reproduced from Bentsen and Olesen (2002) by permission of Elsevier

exceed 100 and preferably be at least 200–300, such a unit is a considerable investment in size and money (Fig. 12.4).

For some species like tilapia and shrimp, a simpler and much cheaper solution is to use hapas instead of covered tanks (Fig. 12.4). Natural spawning and mating may be practised by introducing one male and one female breeder into a hapa.

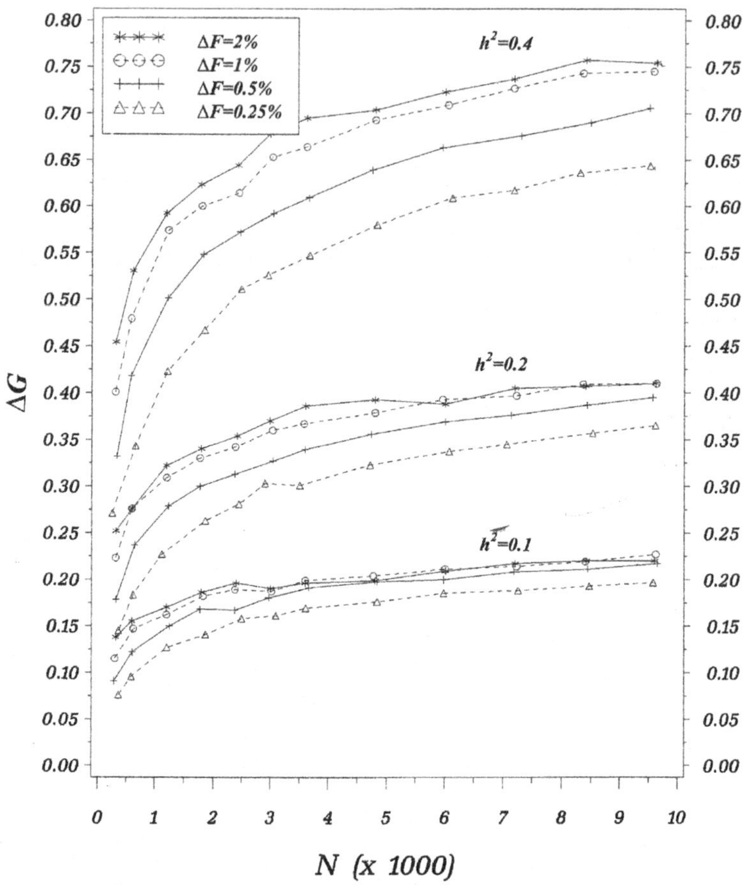

Fig. 12.2 Genetic gain per generation (ΔG) for optimum schemes with different population sizes (N), heritabilities (h^2) and restrictions on the rate of inbreeding (ΔF). In all cases, each sire fertilises eggs from two dams. Reproduced from Gjerde et al (1996) by permission of Elsevier

The fry can be maintained in the hapas until they reach tagging size. To produce half-sibs, the male may be transferred to a second female after successfully mating with the first one. Alternatively, artificial stripping of eggs and sperm can be performed, with fertilised eggs from each family being placed into separate hatchery containers. When fry are ready to commence feeding, each family can be then placed into separate pond compartments. Such a strategy has been successfully implemented for species such as rohu carp (Fig. 12.4). In any large-scale breeding program utilising both family and individual selection and artificial spawning, there are a number of key steps that are described in more detail in the following section.

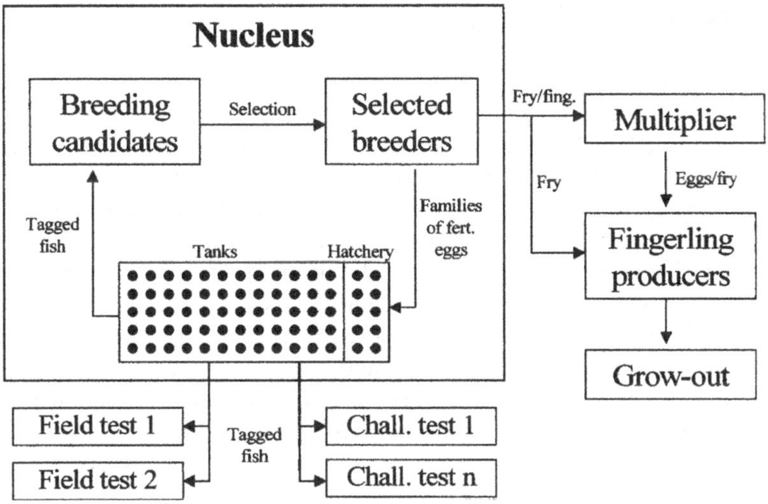

Fig. 12.3 The main elements of a full-scale breeding program in fish. Reproduced from Gjerde (2005b) by permission of Springer

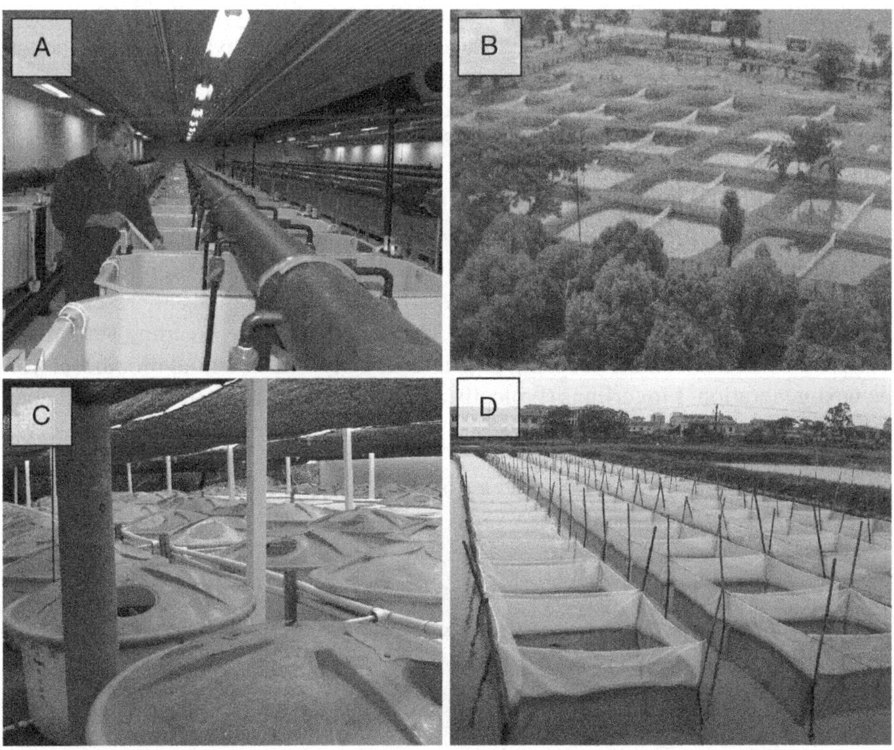

Fig. 12.4 Systems for separate rearing of fry from different families until tagging size. (**A**) tanks for salmon (photo Frode Nerland), (**B**) ponds for carp (photo Bjarne Gjerde), (**C**) tanks for shrimp (photo Morten Rye) and (**D**) hapas for tilapia (photo Morten Rye)

12.3.1 Mating and Hatching

The compelling arguments for the production of both full- and half-sib families are described in earlier chapters. In practice, milt from one male should fertilise eggs from two to three females to produce full- and half-sibs. A sample of these eggs, approximately 500 from each full-sib group is a reasonable number, are then placed into a hatching tray. As start-feeding approaches, the alevins are transferred to a tank, hapa or pond, depending on the species and facilities available, for rearing until the fingerlings can be physically tagged. It is important to standardise the environmental conditions as much as possible during this period, to minimise common environmental effects. Of particularly importance is to use the same water quality and quantity, the same feed quality and quantity, and particular care must be taken when handling fry and cleaning tanks or ponds. As the amount of variation in rearing conditions is reduced, so are the common environmental effects.

12.3.2 Tagging

As soon as the fish reach a sufficient size to allow tagging (typically 5–10 grams), samples from each family are tagged. A wide range of tags are available, some better suited to certain species than others (Fig. 12.5). For fish, the most common and effective tag type in use today is the electronic passive integrated transponder (PIT) tag that is inserted into the body cavity. For shrimp, the visible implant elastomer (VIE) is an effective tag that can be used on larvae as small as 1 g in size (Godin et al. 1996). For molluscs, small tags (such as bee tags) glued to the surface of the shell are effective. In New Zealand, a new method for identifying mussels based on etching an ID code onto the shell has been developed by the Cawthron Institute.

From the animals tagged in each family, around 100 individuals are reserved for communal rearing at the breeding station. They are held at the breeding station for recording of all traits in the breeding goal and also become potential broodstock for the next generation. Fingerlings (typically around 15–25 g) to be used for challenge

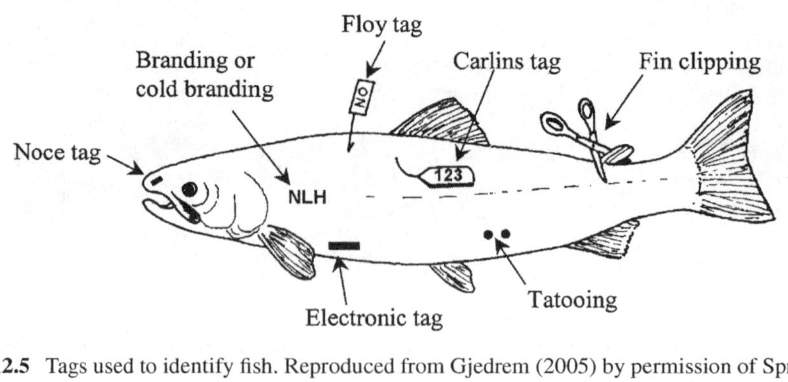

Fig. 12.5 Tags used to identify fish. Reproduced from Gjedrem (2005) by permission of Springer

testing for different diseases are also tagged at the same time. Additional individuals from each family (suggested to be around 50) are tagged to be reared in the field, typically at commercial farms. Breeding programs for fish species typically utilise two to three test stations in order to evaluate families under practical farming conditions.

12.3.3 Recording During Grow-Out

During the grow-out period, some records are taken at the breeding station as well as at the field test stations. The most common traits recorded are age of sexual maturation and mortality (and its cause if known). When the animals approach market size, final records are taken of all animals at test stations and sent to the breeding station. Based on the available data from the test stations, challenge tests and information taken at the breeding station, a preliminary family index can be estimated. When family indexes are available, the harvesting and recording of animals at the breeding station take place.

12.3.4 Harvest and Pre-selection of Broodstock at Breeding Stations

The final recording of individuals takes place during the harvest of animals at the breeding station. Typically, the following measurements are taken from anaesthetised animals:

- Body weight and length
- Sex
- Body shape and abnormalities (if any)
- External colour and other morphometric traits.

This information, together with a ranking list of all families, enables pre-selection of broodstock to be performed. In practice, this could be carried out as follows.

- Males with body weight more two standard deviations above average and belonging to one of the 10 highest ranking families are pre-selected
- Females with body weight higher than one standard deviation and belonging to one of the 20 families with the highest rankings are pre-selected.

At the same time, 15–20 fish from each family are sampled for measuring product quality.

12.3.5 Final Selection of Broodstock

When all records are gathered from the breeding station, field test stations, challenge tests and product quality evaluations, individual indexes are calculated for all pre-selected broodstock based of individual performance together with all records of full- and half-sibs. The final selection of broodstock then takes place, and a mating list constructed, taking into account the relationship between individuals and avoiding the mating of close relatives.

Selected females are usually large, yielding a large number of eggs, and the selected males produce a large amount of milt over several weeks. Since only a few hundred eggs are needed to produce families for testing in the nucleus, the remaining eggs and milt are taken care of and used for other purposes.

12.3.6 Genetic Markers for Parentage Assignment

An alternative to physical tagging is to infer parentage of animals retrospectively based on genetic markers such as microsatellites. The biggest advantage of such a strategy is that there will be no common environmental effect which may occur when families have been reared separately until tagging. Furthermore, it is possible to extend the number of broodstock and families tested in each generation without additional investments.

There are some disadvantages to this strategy however. Pooling individuals at fertilization will lead to variable number of animals per family available for recording as well as for selection due to differences in survival, Table 8.1. In a simulation study, Gjerde (2005b) found that only 52 and 85 out of 100 families had 10 or more fish per family when a random sample of 1,000 and 2,000 individuals were genotyped, respectively. Thus, a large number of animals must be genotyped to get information of all families, and if traits are recorded at different times the animals need be retyped each time. Breeding candidates need to be physically tagged in any case in order to be able to easily trace the candidate at time of mating. Nevertheless, the continual reduction in genotyping costs means that strategies using markers may become increasingly attractive in certain situations.

12.4 Test Stations

The purpose of field test stations is to ensure that the families are tested under practical rearing conditions. It is possible that the environmental conditions at the breeding station may not be representative of the real farming conditions for the species in question. Performance data from different commercial environments ensures that selected animals are robust to different environmental conditions. Although GxE interactions appear to be insignificant for most species even when there is a large degree of variation in farming conditions, testing animals at different farms represents an 'insurance policy' in the cases where such interactions are present. Test stations also serve as reserves for the breeding station in the event of accidents and diseases. As discussed earlier, test stations can be part of a stepwise recording for selection of breeding candidates.

Some records that can be taken at test stations and transferred to the breeding station include the following:

- Frequency of early sexual maturation
- Body weight and length at harvest
- Mortality including weight and date
- Product quality, measured on a given number of animals from each family.

When records are taken prior to corresponding recordings at the breeding station, it is possible to use the data to estimate family indexes for the pre-selection of broodstock at the breeding station.

12.5 Production of Special Lines

Well-designed dissemination schemes may offer the potential for product differentiation to suit particular requirements. While multi-trait selection in the breeding nucleus is generally focused on a long-term breeding goal, it is possible for the breeding station to offer selected lines for short-term priorities, placing emphasis on traits of immediate importance.

In practice, this could be the result of some farmers selling to markets that have special requirements for a product with particular characteristics. To meet these requests, the breeding station may allow some producers to define their own short-term breeding goal. The most interesting selection lines are likely to focus on one or more of the following traits: fast growth rate, resistance to specific diseases, product quality traits and age at sexual maturation. At the breeding station, these lines may be produced by selecting a few males and a few females that have extremely high breeding values for the trait or traits in question, and mating them to produce progeny for these special lines. Development of special lines can be achieved without affecting the selection procedure in the nucleus, where continuous focus on the overall breeding goal is given the highest priority. An example of different selection procedures in a nucleus and the development of special lines is summarised below.

Continuous selection in the nucleus:

- Selection index including: Body weight, disease resistance, product quality, other traits in the breeding goal.

Selection in specially produced lines:

- Index including: body weight
- Index including: One or several diseases
- Index including: One or several quality traits.

12.6 Dissemination of Genetic Gains

In order to maximise the benefit of a breeding program, the genetic improvement should reach the farmers with minimum time lag. Full benefit from the investment placed in the breeding program requires efficient systems for transferring the genetic gain obtained in the breeding nucleus to the farmers. The dissemination strategy has major impact on the benefit/cost ratio of a breeding program. The transfer of genetic gain from nucleus to industry can follow two main paths:

- Direct transfer from the nucleus
- Transfer via multipliers.

A great advantage that aquatic animals have when it comes to dissemination, particularly marine species, is their extremely high fecundity. Females often produce thousands or even millions of eggs, in stark contrast to terrestrial livestock species. Furthermore, for most species it is possible to transport eggs, sperm, alevins and fingerlings as well as older individuals over long distance in small containers, even by air freight. In practice, the dissemination of eggs and fry is the preferred method for distributing genetic gain to the fish farmers.

The transport of live animals from one area to another does, however, involve some risk of disease transfer. In this regard, eggs offer the advantage that they can be disinfected by both the seller and buyer. Nevertheless, there remains the potential of vertical transmission of disease to offspring, although this has not been documented. Figure 12.6 depicts the various ways genetically improved material can be disseminated to farmers. The numbers and scale of the units depicted will depend on the size of the industry, breeding program, and uptake of genetically improved material.

12.6.1 Direct Dissemination from Nucleus

The high fecundity in most aquatic species allows a strong selection of broodstock. In a family based breeding program, the testing of 200–500 full-sib families per generation only requires a few hundred eggs from each female. The excess eggs produced by the female broodstock, likely to number in the thousands or millions for some marine species, could be sold to the industry directly. In addition, the breeding station has the possibility to select broodstock with lower breeding values than those selected for the production of families in the nucleus. The progeny of the 'next best broodstock' will still in all likelihood produce high quality progeny. Therefore, it is possible for a breeding station to transfer genetic gain in the form of high quality eggs or fry directly to the industry.

If the selected broodstock kept at the nucleus are used for dissemination, there will be no time lag between the time that genetic gain is obtained in the nucleus, and when it is received by the end users. However, when eggs or fry from 'next

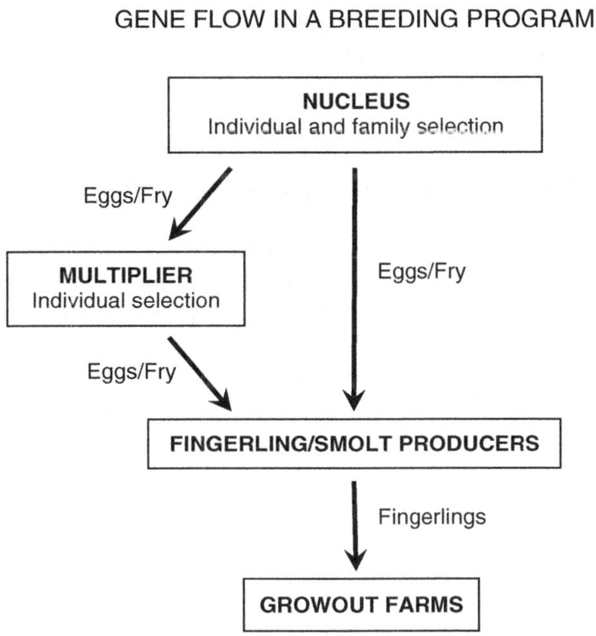

Fig. 12.6 Alternative routes of genetically improved material from a fish breeding nucleus to farmers

best breeders' are used for dissemination, the time lag will be proportional to the breeding value of the actual broodstock used, compared to the breeding values of the selected broodstock.

In a stochastic simulation study, Skagemo et al. (2008) investigated the effect of the selection of broodstock for dissemination of eggs or fry to the industry. In the nucleus, 100 families were tested in each generation and at time of selection there were 2,000 females and 2,000 males available. Random selection of 100 parental pairs was compared with truncation selection of 100 and 10 parental pairs. Higher profit was calculated for all selection schemes compared to random selection, and profit increased with increasing selection intensity. Compared to random selection of nucleus breeders for dissemination purposes, optimising the selection of parents for production of eggs or fry to the grow-out farmers could lead to an additional response corresponding to approximately 1.5 generations of selection in the nucleus. From an economic perspective, the additional profit obtained through applying selection was estimated to be € 250,000 for a medium sized cod farm in Europe marketing 350,000 fish annually with value of 0.5€ per standard deviation in slaughter weight.

The optimal strategy for the dissemination of genetic improvement from the nucleus to the industry depends primarily on the fecundity of the species in question. The higher the fecundity of a species, the larger the capacity of the nucleus to transfer genetic gain directly to the industry. For species with relatively low fecundity, such as tilapia and salmonids, it is not easy for the nucleus to produce sufficient volumes of eggs or fry to satisfy market demands. If the breeding station does not have enough capacity to meet demands, the most common alternative is to establish multipliers.

12.6.2 Dissemination from Multipliers

When a breeding station does not have the capacity to supply enough eggs or fry to meet the market demand, multiplier stations are required. A multiplier can simply be a commercial farm with a hatchery, tasked with receiving eggs or fry from the nucleus, and growing them to produce broodstock. After applying strong individual selection for growth rate, these 'multiplied' broodstock are then used to produce eggs or fry to be sold to the industry. The main elements in the agreement between the breeding station and multipliers will usually be:

- Breeding stations shall deliver eggs/fry of broodstock from the last generation with highest possible breeding values
- Multipliers shall only produce broodstock from the transferred eggs/fry
- Broodstock used for production of eggs/fry at the multiplier must be selected strongly for growth rate
- The breeding station and multiplier shall share the market for eggs/fry of the strain in question without competition.

There are several strategies for selection of broodstock at the multiplier level:

- Individual selection of males and females for growth rate
- Individual selection of females for growth rate and use of milt of males from the nucleus selected for all traits in the breeding goal
- Selection within lines produced by the nucleus
- Selection based on family indexes estimated at the nucleus, when the multiplier receives tagged fingerlings.

Skagemo et al. (2008) studied the optimisation of broodstock selection at multipliers in breeding programs. A conclusion reached was that the selection of parents for eggs or fry that are transferred from the nucleus to the multipliers has a much larger effect on the genetic value than selection of broodstock at the multipliers. However, strong potential for gain through the selection of broodstock at the multipliers was also identified.

Usually, multipliers can only apply individual selection for growth rate on male and female broodstock. However, technological advances in instrumentation for measuring other traits like fat and colour on live breeding candidates may result in the possibility for individual selection for additional traits (Folkestad et al. 2008). Individual selection for both growth rate and quality traits will naturally result in a higher genetic value of eggs or fry transferred from multipliers to the industry.

Multipliers can alternatively fertilise selected females with milt from highly selected males provided directly from the nucleus. These males may be highly selected for a selection index or for particular traits of special interest to certain industry members. This alternative has some advantages compared with using broodstock individually selected for growth rate within the multiplier. If a multiplier plans to use milt from the nucleus, they may slaughter all males at an early stage and thus reduce costs and increase testing capacity.

Multipliers can exploit special lines created in the breeding nucleus through the production of broodstock from eggs or fry received from these lines in the nucleus. The genetic merit of the line produced at the multiplier can be increased by using the milt of selected males in the nucleus. Special lines for growth rate are of particular interest for multipliers since both males and females can be selected.

In a large-scale breeding program, multipliers will play a central role in the dissemination of genetic gain to the industry. Multipliers will generally receive eggs or fry equivalent to the most superior families in the nucleus. The main difference is that selection of broodstock at the nucleus is generally based on a total index value while multipliers usually select on growth rate only. However, since body weight has a relatively high heritability, the difference will be relatively small if the same selection intensity is applied at the multiplier and nucleus levels. This enables the multipliers to disseminate most of the genetic gain obtained in the nucleus to the industry with minimal time lag. This is very important, since long-term genetic improvement is fundamentally the result of the magnitude of genetic improvement obtained each year.

12.7 Breeding Programs for New Species

History has shown that it is not easy to start farming a new aquatic species. A large number of issues concerning basic husbandry and rearing conditions must be resolved. Diseases will occur and treatments must be developed, and suitable feeds have to be produced. The wild animals will not be adapted to captivity, and show signs of stress, although domestication will take place immediately from the first generation produced in captivity. Breeding programs have the ability to accelerate the domestication process and rapidly increase productivity.

A basic requirement for implementing a breeding program in a new species is that there is knowledge and control of the entire life-cycle. However, it is likely that quantitative genetic parameters will be lacking, demanding some initial research before full implementation of a breeding program can occur. Important tasks for generating the basic foundation for a breeding program include:

- The study of genetic parameters for important production traits in the most promising strains or populations of the species
- The estimation of phenotypic and genetic parameters for all traits in the breeding goal, particularly phenotypic and genetic variation, heritability, genetic and phenotypic correlations

- Quantification of the degree of heterosis that occurs when crossing strains, information that can help decide whether a selection or a combined selection and crossbreeding scheme should be applied
- The investigation of potential GxE interactions
- A clear definition of the breeding goal
- The production of a base population, consisting of crosses among the most promising strains in order to establish a broad genetic base
- The development of selection indexes
- The planning and development of necessary infrastructure to run the breeding program
- The planning of a strategy to estimate selection response
- The planning of a strategy for the dissemination of improved stock.

To develop a breeding program for a new species, all these points are important and need to be considered. A key factor influencing the structure of the breeding program is the fecundity of the species in question. As most aquatic species are very fertile, established and successful breeding programs such as that used for Atlantic salmon in Norway, can serve as an excellent model. For multiple-trait breeding programs, family testing and selection will be central. When the heterosis effect is considerable, a breeding program based on crossbreeding only is inefficient and should always be combined with family and within-family selection. Progeny testing may be a viable strategy for multiple spawners, but is not widely used in aquaculture given the major disadvantages inherent in this strategy.

The importance of obtaining accurate estimates of basic genetic parameters before commencement of the breeding program may necessitate collaboration with academic personnel experienced in quantitative genetic theory. In the initial phase of the breeding program, the productivity of the species will be low, and therefore there may be justification for governmental authorities to contribute financially to R&D and other critical tasks related to the initiation of the program. However, farmers and the industry as a whole who want to initiate production of the species should naturally also make a major contribution to the development.

In countries and regions where farmers do not have previous experience with selective breeding and have limited facilities, the starting phase should be simple, with a primary aim of convincing the farmers of the benefit of genetically improved stock.

Some basic, universal strategies can be applied when a new breeding program is initiated. A fundamental initial step is the crossing of local strains and available improved stocks, using many broodstock from each strain in order to secure a broad genetic base. For the two first generations, selection should mainly focus on fast growth rate in order to demonstrate and document rapid response. Appropriate control groups will show the magnitude of selection response obtained and demonstrate the economic benefit of the breeding program. Once this is established, longer-term goals can be considered and the program scaled up appropriately.

Chapter 13
Undesirable Side Effects in Breeding Programs

13.1 Introduction

In general, breeding programs have positive effects on welfare and productivity. However, apart from the desired effects of genetic gain, breeding programs may produce various types of unfavourable side effects. According to Rauw et al. (1998), animals selected for high productivity seem to be more at risk of behavioural, physiological and immunological problems. Such undesirable side effects are largely caused by genetic correlations between traits included in the breeding goal and traits not targeted for by direct selection. Therefore, it is extremely important to monitor as many traits as possible, particularly those related to fitness, even if they are not included in the breeding goal. The purpose is to observe as early as possible any adverse effects caused by selection, allowing precautions to be taken before damage occurs.

Problems related to correlated selection responses may be very complex. When a species is domesticated, it must cope with a completely new growing environment, new type of food, limited space, crowded conditions, and possibly increased stress levels. This stress often results in a marked impact on behaviour and development. This highlights that biological changes are not necessarily only caused by selection procedures, but also by environmental factors.

13.2 Correlated Effects

In Chapter 4, Fig. 4.8 was used to highlight relationships between traits. The genetic correlations are of particular interest when discussing undesirable effects of breeding programs. An unfavourable genetic correlation between a trait in the breeding goal and a trait that was not considered important enough to be included may lead to undesired changes in this trait through selection. There are several examples of such changes in livestock.

Rauw et al. (1998) reviewed the literature on undesirable side effects of selection for high production efficiency in terrestrial livestock. They documented examples of

T. Gjedrem, M. Baranski, *Selective Breeding in Aquaculture: An Introduction*, Reviews: Methods and Technologies in Fish Biology and Fisheries 10,
DOI 10.1007/978-90-481-2773-3_13,

over 100 references of undesirable correlated effects of selection for high production efficiency, with respect to metabolic, reproduction and health traits in broilers, pigs, and dairy cattle. The most striking example was found in broiler chickens where an increasing incidence of heart failure syndrome and leg problems had occurred following intense selection for body weight at a certain age. From several investigations in dairy cattle, an antagonistic relationship was found between high production of milk and several fertility traits, and it was concluded that: 'In general, high producing cows were bred later, showed more days open, had a longer calving interval, a lower rate of non-return at 56 days, and required more services per conception than low producing cows'.

In broilers, Havenstein et al. (1994) reported more than a fourfold increase in mortality at 42 days of age for a strongly selected line, compared to the original population from 1957. Most of the deaths were associated with sudden death, ascites and leg problems. Similarly, in pigs, Sather (1987) experimentally found significantly more leg weakness in the foreleg and rear leg than non-selected control boars.

These examples highlight that high selection intensity for production traits may lead to loss of homeostatic balance causing undesirable side effects for fitness traits such as tolerance to diseases, reproduction and welfare. The biological explanation is discussed by means of the Resource Allocation Theory (Beilharz et al. 1993). When resources are limited for the animal, a compromise has to be found on how to partition available resources among traits. When artificial selection is focused on high production, the animals will allocate resources to the traits defined in the breeding goal and less resources will be left to other demands, like coping with stressors.

Most of the estimated genetic correlations between disease resistance and production traits in fish and shellfish are low and positive (Table 5.1), but there are also some negative correlations between production traits and diseases caused by viruses. Henryon et al. (2002) estimated relatively low negative genetic correlations between the predicted breeding values for VHS resistance and the predicted breeding values for growth rate and feed conversion efficiency ($r_G = -0.01$ to -0.22) within a rainbow trout population. In shrimp, Gitterle et al. (2005) found a relatively high negative genetic correlation between harvest body weight and resistance to white spot syndrome virus (WSSV) while Fjalestad et al. (1997) estimated a low genetic correlation between body weight and taura syndrome virus (TSV). The implication is clear. Breeding programs for fish that concentrate solely on the improvement of production traits (i.e., growth rate and feed conversion efficiency) are likely to have adverse effects on some health traits.

As soon as negative changes are observed, the breeding goal should be redefined and the affected traits should be included in the breeding goal. Negative side effects in aquatic species are most likely to occur in fitness traits like gonad quality and survival in early life stages, in addition to internal changes like size and form of the heart.

13.3 Breeding Goal May Change

Traits like growth rate and disease resistance are traits likely to be included in the breeding goal on a stable and long-term basis, since rapid growth and high survival have high economic importance and are universally advantageous from a production point of view.

For species that become sexually mature before they reach market size, an important part of the breeding goal will be to reduce the frequency of early sexual maturation. However, when response to selection for increased growth rate becomes high, early sexual maturation may change from a negative effect to becoming an advantage.

Changes in the breeding goal for product quality traits are more likely to occur than for other traits. A good example can be seen in pigs. In former times it was considered to be desirable to have a large amount of back fat on the carcass, while low back fat thickness is now given priority. The same may happen for fish species with high fat levels, since the market may want a leaner product in the future.

New traits may become important economically. For many traits, it is extremely difficult or simply not possible for measurements to be taken on live individuals, with feed conversion efficiency being a classic example. This may change in the future with the development of new methods for direct or indirect measurements of a trait. Improved measuring techniques may become possible for stress, behaviour, welfare, product quality, diseases and characters genetically correlated with central traits included in the breeding goal.

13.4 Disease Prevention

For aquatic species, most breeding programs are centralised. Breeding companies together with a number of licensed multipliers may cover a large market spread over a broad geographic range. If an infectious disease enters this system, it may be rapidly spread throughout the industry. Therefore, high hygiene standards throughout the whole production and testing system must be given high priority. A thorough and continuous disease monitoring program must be applied at both the breeding station as well as the multipliers. The production of healthy animals for the industry is a vital part of any breeding program.

Both fish and shellfish can be transported over long distances, and relatively easily be treated with drugs to kill bacteria, viruses and parasites. Eggs therefore are an efficient means of distribution of genetically improved material, especially considering the low incidence of vertical transmission of diseases.

It has been argued that selection for disease resistance may result in an increased proportion of carriers that could spread the disease to their surroundings. However, evidence suggests that 'susceptible' animals are more likely to be infected in the first place and therefore pose a larger risk for spreading diseases.

13.5 Genotype–Environment Interaction

Genotype–environment (G×E) interactions were discussed in detail in Chapter 10. Although the incidence of significant interactions generally appears to be low in aquatic species, there is sufficient evidence to show that this is relatively unpredictable. It is therefore important that interaction effects are studied during the early phase of breeding programs, covering the range of environments that the improved stock are likely to be raised in. If G×E interactions are important, establishing two or more separate programs must be considered.

Over time, the environmental conditions in the industry may change, new technology may be introduced, and new areas may be taken into production. Such changes may be so dramatic that a different breeding goal may become necessary. There is also the possibility that a selected strain may actually become more sensitive to environmental variation warranting further investigation of possible G×E interactions.

13.6 Increase of Inbreeding

In a closed breeding population, it is impossible to completely avoid inbreeding. The larger the population is and the higher the effective number of breeders (Ne) is, the slower inbreeding will accumulate. The risk of high inbreeding is particularly high when animals are untagged.

A rapid accumulation of inbreeding must be taken seriously since it will result in reduced performance and detrimental genetic effects. It is well documented that growth rate will be reduced and mortality may increase by 1–9% per 10% increase in the coefficient of inbreeding (F) (Fjalestad 2005). Over time these changes may become a serious handicap for the stock.

In addition to reducing the performance of animals, inbreeding will reduce the genetic variance, which will result in a reduced response to selection. Several selection experiments have shown that genetic gain reached a plateau after 20–50 generations of selection, such as selection for thorax length in *Drophila melanogaster* (Robertson 1955) and selection for six week old body weight in mice (Roberts 1966). The major factor responsible for the plateau in these cases was the use of relatively few broodstock.

13.7 Conclusion

It is now well documented that selective breeding is a powerful tool to improve production traits in aquatic species. Response to selection is several times faster for aquatic species compared with terrestrial livestock species. The main reason for this difference is the high fecundity of fish and shellfish, which facilitates high selection intensity. However, at the same time, this intense selection may increase susceptibility to undesirable side effects. It is therefore essential that each breeding program takes necessary steps to document negative effects as early as possible and make changes to eliminate the problems before they become harmful.

It is also important to emphasise that breeding programs should have broad breeding goals in order to develop robust animals for a wide range of environmental conditions. Evidence from terrestrial species shows that strong selection pressure on one production trait will produce animals with little tolerance to variation in environmental conditions.

Since effects of selection appear in future generations, any changes in consumer preference should be identified as early as possible, so the breeding goal can be altered and minimal delay occurs between this change and dissemination to industry.

One of the biggest challenges for a breeding company is to avoid the spread of infectious diseases. Rigorous protection and control regimes should be implemented to prevent production losses from outbreaks of disease, and prevent spread to industry and the natural environment.

Chapter 14
Biotechnology in Breeding Programs

14.1 Introduction

Modern biotechnology has had a dramatic impact in the field of human genetics, and is becoming an increasingly important part of breeding programs in terrestrial livestock species. The application of molecular genetics in aquaculture species has tended to lag behind livestock, given the development of industries for livestock species tend to be far in advance of most aquaculture species, and therefore most of the basic questions regarding husbandry and breeding have been resolved. In addition, the genomic resources for species such as cattle and pigs are far in advance of even the most well developed aquaculture species (salmonids). However, given the rapid advances in DNA sequencing and genotyping technology that are now occurring, these resources in aquatic species will develop rapidly and potentially enable similar molecular genetic approaches to be taken to improve genetic gain in selective breeding programs.

14.1.1 DNA Markers

Novel genetic technologies involving the use of DNA-based tools are under development for a range of aquaculture species (Davis and Hetzel 2000). A wide range of genetic markers have been developed for application in genetic improvement and management, each having its particular advantages depending on the application. These markers are broadly classified into two categories: type I are markers associated with genes of known function, while type II markers are associated with anonymous genomic segments (Liu and Cordes 2004). To date, the most commonly used DNA markers for application in fisheries and aquaculture have been allozymes (Hamm and Burton 2000), restriction fragment length polymorphism (RFLP) (Lander and Botstein 1989), random amplified polymorphic DNA (RAPD) (Huang et al. 2000), amplified fragment length polymorphism (AFLP) (Vos et al. 1995; Li and Guo 2004), microsatellites (Schlötterer 2000; Hara and Sekino 2005), sequencing of both nuclear and mitochondrial DNA (Hamm and Burton 2000) and

T. Gjedrem, M. Baranski, *Selective Breeding in Aquaculture: An Introduction*,
Reviews: Methods and Technologies in Fish Biology and Fisheries 10,
DOI 10.1007/978-90-481-2773-3_14, © Springer Science+Business Media B.V. 2009

single nucleotide polymorphisms (SNP) (Moen et al. 2008b). Microsatellites and SNPs are the most commonly used markers today, and are described in the following sections.

14.1.2 Microsatellites

Microsatellites consist of local repetition of simple sequence motifs that range in size from one to six base pairs (Tautz 1989) and are inherited in a Mendelian fashion. Although microsatellites are generally considered to be evolutionary neutral DNA markers, they have been implicated in functional significance through critical tests in various biological phenomena (Rocha et al. 2002). Compared to other popular DNA marker systems, microsatellites offer the advantages of being highly repeatable, easily and accurately scored, codominant and having the highest polymorphic information content (PIC) of any DNA marker (Liu and Cordes 2004). Some disadvantages of microsatellites include the fact that they are relatively expensive and time consuming to develop since prior sequence knowledge is required for primer design, and transferability is generally limited to within genera at best (Fig. 14.1).

14.1.3 Single Nucleotide Polymorphisms (SNPs)

Recently, single nucleotide polymorphism markers (SNPs) have become the marker of choice for studies related to gene mapping and selection. SNPs are the most abundant type of polymorphism in the genome and consist of a DNA sequence vari-

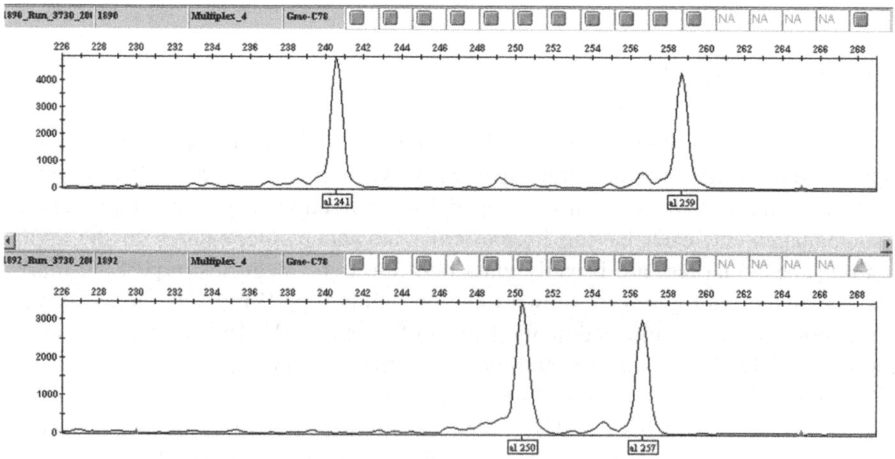

Fig. 14.1 Example of two microsatellite genotypes. Two heterozygous individuals are represented, with alleles of size 241, 259 and 250, 257

ation occurring when a single nucleotide in the genome (or other shared sequence) differs at a given site. SNPs are found in both non-coding and coding DNA. Coding SNPs in a particular gene can actually be a missense mutation having a large effect on phenotypic variation. SNPs are however, less informative than microsatellites, and it is estimated that around five SNPs provide equivalent information to a single microsatellite. This disadvantage has been dramatically offset by the development of high-throughput SNP genotyping technologies, such as those produced by Illumina™ and Affymetrix™. SNP genotyping arrays containing over one million SNPs in a single assay are now available in humans, and genotyping costs of SNPs continue to decrease.

14.2 Linkage Maps

Genetic linkage maps are a major step towards the use of genomic information in livestock improvement and are designed to show the genetic distances and orders of markers in the genome. Whole genomes of some of the major livestock and model animals have been mapped with several thousand markers (Knapik et al. 1998; Tong and Chu 2002; Ihara et al. 2004). This makes it possible to choose a set of linked markers to mark any segment of the genome or many segments to mark the whole genome. Once loci of particular DNA sequences have been mapped in one species, the information is highly useful in genome mapping in another. The mapping of livestock species has been greatly enhanced by the completion of genome sequences of among others mouse (Waterston et al. 2002) and humans (Venter et al. 2001). An essential complement to genetic linkage maps are physical maps upon which stretches of known DNA sequence are placed (e.g. microsatellites, genomic library clones, ESTs etc.). Unfortunately, even the karyotypes of many aquaculture species have not been well defined (Tong and Chu 2002) and as a result the extent of genome coverage with marker maps cannot be precisely defined.

The most widely used method of producing linkage maps is the transformation of the observed recombination fractions into an additive map distance using a mapping function. Mapping functions account for the fact that double and other even numbers of recombinants cannot be observed. One of the commonly used mapping functions was derived by Haldane (1919), and assumes crossovers occur randomly and independently over the entire chromosome. An alternative, the Kosambi (1944) mapping function, accounts for interference, where the presence of a crossover in one region affects the frequency of crossovers in other regions. Recombination and map distances often vary markedly between sexes, this is particularly evident in salmonids where male recombination is significantly lower than female recombination (Sakamoto et al. 2000; Moen et al. 2004c). This has also been observed in molluscs, where recombination rates and marker order have been observed to differ between male and female parents and among parents of the same sex (Hubert and Hedgecock 2004).

The units of map distance estimated by mapping functions are Morgans (M) and centimorgans (cM). One centimorgan is equal to a 1% chance that a marker at one

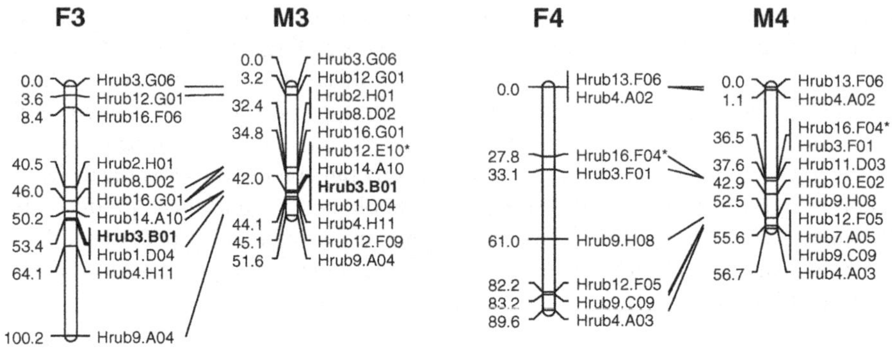

Fig. 14.2 Two linkage groups for the blacklip abalone (*Haliotis rubra*). Note the larger amount of recombination in the female map (F)

Table 14.1 A selection of aquaculture species with linkage maps developed

Species	Reference
Pacific oyster	Hubert and Hedgecock (2004), Li and Guo (2004)
Channel catfish	Waldbieser et al. (2001)
Tilapia	Kocher et al. (1998), Lee et al. (2005)
Eastern oyster	Yu and Guo (2003)
Japanese flounder	Coimbra et al. (2003)
Yellowtail	Ohara et al. (2005)
Common carp	Sun and Liang (2004)
Zhikong scallop	Wang et al. (2005)
Atlantic salmon	Moen et al. (2004c), Gilbey et al. (2004), Moen et al. (2008b)
Rainbow trout	Sakamoto et al. (2000), Nichols et al. (2003b)
Brown trout	Gharbi et al. (2006)
Black tiger shrimp	Wilson et al. (2002), Staelens et al. (2008)
Kuruma prawn	Moore et al. (1999)
European sea bass	Chistiakov et al. (2005a)
Blacklip abalone	Baranski et al. (2006)

genetic locus on a chromosome will be separated from a marker at a second locus due to crossing over in a single generation. Figure 14.2 shows two linkage groups from a linkage map of the blacklip abalone constructed with the Kosambi mapping function.

Compared to many livestock and model organisms that have high density maps of many thousands of markers, the relatively few linkage maps that are available in aquatic species tend to be of much lower density (Table 14.1). Aquatic species also suffer from a lack of inbred lines from which the ideal mapping families can be derived, but offer the advantage of generally high fecundity (i.e. large family sizes), so extensive map data can be obtained from relatively few parents. The high natural heterozygosity of many aquatic species means that natural populations are often good mapping resources (Moen et al. 2004c; Wang et al. 2005).

14.3 Quantitative Trait Loci (QTL)

The development of genetic markers linked to genes affecting commercially important traits has been the focus of a number of genetic improvement efforts. The effect on phenotype can be manifested in two ways: (1) as single genes inherited in a Mendelian fashion that essentially control an observed phenotype and; (2) as many genes of small or large effect on a trait that is quantitative in nature (Davis and Hetzel 2000). Localised regions of the genome containing genes affecting quantitative traits are known as quantitative trait loci (QTL). QTL are of particular interest in aquaculture as most traits of commercial interest show continuous or quantitative variation. Knowledge of linkage between molecular genetic markers and particular QTL enables the use of marker assisted selection (MAS) (Poompuang and Hallerman 1997). QTL detection has the ability to deliver major gains for traits that are otherwise difficult or expensive to measure (such as food conversion efficiency and disease resistance), can only be measured after the normal period of selection (such as reproductive characteristics) or can only be measured in a destructive manner (such as sacrificing potential broodstock for flesh quality traits). QTL studies in aquaculture species have largely focused on salmonids, owing to their prevalence in aquaculture worldwide, maturity of culture systems and high economic value (Table 14.2).

Table 14.2 A selection of QTL studies in aquaculture species

Species	Trait	Reference
Rainbow trout	Upper thermal tolerance	Jackson et al. (1998), Perry et al. (2005)
	Spawning time	Danzmann et al. (1999), Sakamoto et al. (1999)
	Embryonic development	Robison et al. (2001), Sundin et al. (2005)
	Disease resistance	Ozaki et al. (2001), Nichols et al. (2003a), Rodriguez et al. (2004)
	Length	Perry et al. (2005)
Tilapia	Cold tolerance	Moen et al. (2004a)
	Body colour	Howe and Kocher (2003)
	Salinity tolerance	Lee (2003)
Channel catfish	Feed conversion efficiency	Karsi et al. (2000)
Atlantic salmon	ISA resistance	Moen et al. (2004b)
	IPN resistance	Houston et al. (2008)
	Body weight and condition factor	Reid et al. (2004)
Arctic charr	Body weight, condition factor, age at sexual maturation	Moghadam et al. (2007)
Common carp	Cold tolerance	Sun and Liang (2004)
Blacklip abalone	Growth rate (body weight)	Baranski et al. (2008)

Recently, a major QTL has been identified for IPN resistance in Atlantic salmon in both Scottish (Houston et al. 2008) and Norwegian populations (Moen et al. 2008a). This QTL explained around 26 and 80% of the phenotypic and genetic variances for IPN resistance, respectively in the Norwegian study.

There are two main approaches for the identification of QTL, the candidate gene approach and the QTL mapping approach.

14.3.1 Candidate Gene Approach

The candidate gene approach assumes that a particular mutation within a gene implicated in the physiology of the trait is responsible for a large amount of the phenotypic variation observed. The candidate gene approach involves the sequencing of the gene, or parts of the gene in particular animals that show variation for the trait. Identified polymorphisms are then tested for association with the phenotypic values. The candidate gene approach has been successful in some cases, one good example is a mutation discovered in the oestrogen receptor locus (ESR) in pigs that had a large effect on litter size (Rothschild et al. 1991). However, this approach has had limited success overall (Aguirre-Hernandez and Sargan 2005) for a number of reasons. Firstly, there is typically a relatively large number of genes that affect a given trait, so many genes must be sequenced and analysed for association in order to find the important mutations. Secondly, the causative mutation may lie in a gene that would not have been suspected as an obvious candidate for this particular trait. In a review of candidate gene studies for canine retinal diseases, most of the results (66.6%) excluded the presence of a mutation in a gene or its coding region, while only 3.4% of the results identified the mutation causing the disease (Aguirre-Hernandez and Sargan 2005).

14.3.2 QTL Mapping Approach

An alternative is the QTL mapping approach, in which chromosome regions associated with variation in phenotypic traits are identified. QTL mapping assumes the actual genes which affect a quantitative trait are not known. Instead, this approach uses neutral DNA markers and looks for associations between allele variation at the marker and variation in quantitative traits. Association between a quantitative trait and genetic markers can be evaluated using single markers or multiple markers. The basic methodology involved in detecting QTL combines marker and phenotypic information to obtain correlations between the inheritance of particular marker alleles and particular phenotypes expressed. The large family sizes typically possible for aquaculture species means that QTL mapping experiments can be performed with high power. A good summary of QTL mapping designs for aquaculture species can be found in Massault et al. (2008).

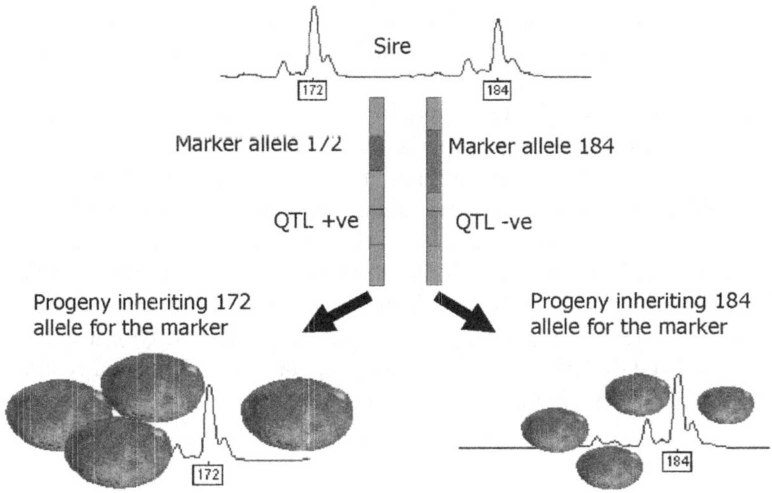

Fig. 14.3 Schematic of marker segregation and co-inheritance with QTL. At this marker, the sire carries the marker alleles 172 and 184. The progeny that inherit the 172 marker allele, linked to the positive QTL allele, tend to be larger than the progeny inheriting the 184 allele, linked to the negative QTL allele

14.3.3 Linkage Analysis

The simplest approach in QTL identification is the use of a t-test or ANOVA to test if the phenotypic means of the progeny groups that inherit different marker alleles are significant for the trait (Kearsey 1998). This approach does not make use of a marker map and cannot provide information about the size or position of the QTL. An analogous approach is to regress the trait value onto the marker genotype (Kearsey 1998). Figure 14.3 illustrates the principle of using the segregation of a marker to infer linkage to a QTL.

If two (or more) markers are jointly used in an analysis, the effect of QTL magnitude and position can be far better separated, and there is more power in detecting a QTL, even if the markers are far apart. A powerful form of QTL analysis utilising multiple markers and map positions is interval mapping (Lander and Botstein 1989). Interval mapping explores the interval between pairs of markers for the presence of QTL, looking at the trait information from each adjacent pair of markers and using this information to infer the likelihood of a QTL being at any given position between them (Kearsey 1998).

14.3.4 Fine Mapping of QTL

Fine mapping describes the identification of markers located closer to the QTL. This is useful since markers identified to be linked to QTL within families can typically be up to 20 cM away from the QTL, so recombination will break up the association relatively rapidly. This means that different marker alleles will be linked to a given QTL allele in different families.

Markers located close to a QTL may be in population-wide linkage disequilibrium (LD) with the QTL. Population-wide LD means that these distance between the marker and QTL is such that the association is not broken down rapidly by recombination and persists across many generations. Therefore the same QTL allele may be linked to the same marker allele across the entire population. A prerequisite of fine-mapping a QTL is a sufficiently dense marker map, something that has been limited in aquaculture species to date. With the advent of large scale marker discovery and high-throughput genotyping technology (e.g. SNP arrays), it is becoming possible to proceed directly to such fine mapping analyses, dramatically improving the efficiency of QTL mapping.

14.3.5 LDLA Mapping

In order to avoid false positives in some situations, it is useful to combine information from across families (LD) and within families (linkage). This type of QTL mapping is referred to as LDLA (linkage disequilibrium linkage analysis). Hayes et al. (2006) modelled the power and accuracy of combined linkage disequilibrium linkage analysis (LDLA) to detect QTL in the commercial population of Atlantic salmon. When 15 half-sib sire families (each sire mated to two dams, each dam with 10 progeny) were sampled from the population for genotyping, it was possible to detect a QTL explaining 10% of the phenotypic variance in 85% of replicates and position this QTL within 3 cM of the true position in 70% of replicates. The results suggest that even with the existing recording structure in commercial salmon breeding programmes, there is considerable power to detect and accurately position QTL using LDLA.

14.3.6 An Example of QTL Mapping to Gene Discovery

A good example of the QTL mapping process that led to the discovery of a very significant mutation is identification of the cattle DGAT1 gene. Commencing with a genome-wide linkage analysis, a region was identified on chromosome 14 that had a large effect on fat percentage (Georges et al. 1995). The confidence region around this QTL was large and potentially contained many genes that could be carrying the underlying mutation. The confidence interval around the QTL was substantially narrowed using linkage disequilibrium mapping (LD) and combined linkage disequilibrium linkage analysis (LDLA) (Riquet et al. 1999; Farnir et al. 2002). As a result of these investigations, the DGAT1 gene was implicated and a single base pair mutation identified that had a substantial effect on milk yield and composition (Grisart et al. 2002).

14.3.7 Strategies to Reduce Genotyping Requirements

QTL mapping experiments can often require thousands of markers and samples to have sufficient power for detection. This means that genotyping requirements, and therefore costs, are also high. Selective genotyping is a strategy to reduce the amount of genotyping required (and therefore cost) for a QTL mapping experiment. In simple terms, this is a method in which the analysis of linkage between markers and QTL is carried out through genotyping of individuals from the high and low phenotypic tails of the trait distribution in the population, rather than all individuals (Darvasi and Soller 1992). However, when multiple traits are of interest (especially those that are uncorrelated), selective genotyping for each trait must be applied to separate samples (selected for each particular trait), and becomes less efficient.

DNA pooling, when used in conjunction with selective genotyping, can reduce genotyping costs of QTL detection by up to two orders of magnitude (Darvasi and Soller 1994). Selective DNA pooling involves the combination of equal amounts of DNA from phenotypically similar individuals to form pools, that are subsequently genotyped with DNA markers (Fig. 14.4). Differences in allele frequencies in the two pools are estimated based on the height or intensity of the signal for each allele in the pool. Markers that show potential linkage to QTL can then be individually genotyped for confirmation of marker-QTL association. Selective DNA pooling has been successfully used to detect QTL in abalone (Baranski et al. 2008), it has effectively been used in mice (Benjamin and Sandra 1996), dairy cattle (Mariasegaram 2004) and humans (Johnson and Griffiths 2005).

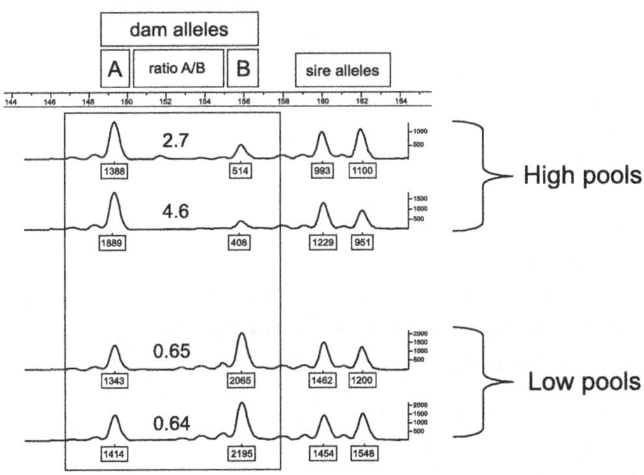

Fig. 14.4 Example of selective DNA pooling results for body weight in abalone. Peak heights for each of the four alleles of a microsatellite marker are shown (in RFUs) below each peak. Two high pools and two low DNA pools are shown along with the ratio of peak heights between dam alleles *A* and *B*, that fast growing abalone tend to inherit allele *A*, while slow growers tend to inherit allele *B*

14.4 Marker Assisted Selection

Marker assisted selection (MAS) is a term used to describe the selective breeding process in which future broodstock are selected based on their genotypes (Liu and Cordes 2004) or on a combination of value estimates made on the basis of marker genotype and phenotypic trait data. To date, this technology has yet to significantly affect the aquaculture industry, however it is becoming a prominent part of many terrestrial livestock breeding programs. The successful implementation of MAS is dependent on a number of factors, primarily: the understanding of the true number of QTL affecting a performance or production trait; the mode of inheritance and relative contribution of the QTL, the linkage and potential interactions of different QTL for the trait and for other traits and the economic importance of each trait (Poompuang and Hallerman 1997).

Given that selection is generally made on a number of different traits, estimated breeding values (EBVs) should be predicted including available phenotypic, pedigree and marker information. Goddard (1992) describes a methodology to achieve this using a bracket of markers surrounding the QTL. The advantage of MAS over non-MAS is approximately proportional to the percentage of the genetic variance accounted for by the marked QTL (Meuwissen and Goddard 1996; Spelman et al. 1999). The key questions then are how many QTL underlie the variation in quantitative traits, and how many of these QTL are necessary to explain the majority of the genetic variance for a typical quantitative trait. Results from powerful genome scans with thousands of SNP markers are beginning to shed more light on these questions.

A summary of the substantial number of gene or marker tests that were being implemented in commercial livestock breeding programs in 2004 can be found in Dekkers (2004). Prior to the advent of DNA markers, selection for individuals genes were implemented based on observable genetic defects and appearance, such as the halothane test as a physical test for the RYR gene, and use of the B-blood group as a physiological LD marker for selection for disease resistance in poultry, which started in the 1960s. Although terrestrial livestock species are far ahead in terms of the application of MAS, a very large QTL identified for IPN resistance in Atlantic salmon has recently been implemented in MAS in Norway, and could increase the rate of genetic improvement for this trait by up to 50% by enabling within-family selection for a disease for the first time.

It is important to note that the application of MAS is not a task that can be performed alone, but must be integrated into all of the other aspects and sources of information in a breeding program (Fig. 14.5).

14.4.1 Types of Marker Assisted Selection

MAS can be carried out within families, once it is established which marker allele is linked to each QTL allele (linkage equilibrium MAS or LE-MAS), across families using markers in population-wide linkage disequilibrium with the QTL (linkage

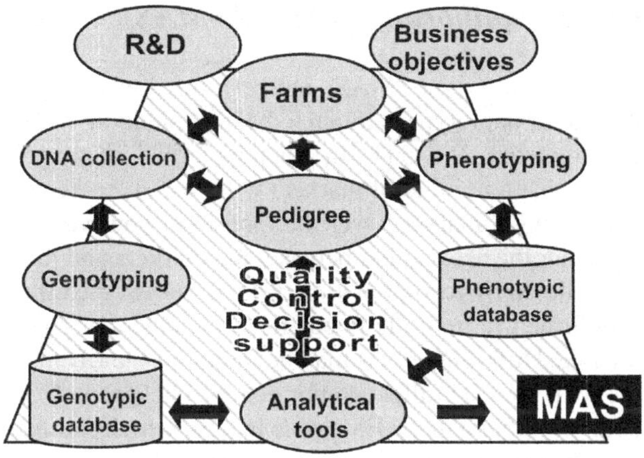

Fig. 14.5. Components of an integrated system for the use of molecular genetic information in breeding programs for marker assisted selection (MAS). Reproduced from Dekkers (2004) by permission of Journal of Animal Science.

disequilibrium MAS or LD-MAS), or using the causative mutation itself (gene assisted selection or GAS).

The three types of marker-assisted selection (GAS, LD-MAS and LE-MAS) differ in their efficiency and ease of implementation. Whereas direct markers LD markers can be used to select across the population because of the consistent association between genotype and phenotype, use of LE markers must allow for different linkage phases between markers and QTL from family to family. Therefore, direct gene markers provide the most powerful tool for selection, followed by LD markers and LE markers.

14.4.2 Gene Assisted Selection (GAS)

Gene-assisted selection (GAS) is the most powerful and straightforward implementation of MAS, and utilises the actual functional mutation or mutations underlying a particular QTL. The effect of the mutation on all of the traits in the breeding goal can be measured, and then included in breeding value estimation as a systematic effect. To implement GAS, only the selection candidates themselves need to be genotyped, however genotype probabilities of ungenotyped individuals should also be estimated. For applications in very distinct populations or differently selected lines, the effect of a functional mutation should first be verified before being universally implemented, as its effect may be influenced by the background genotype. This has been demonstrated for the double muscling locus in beef cattle.

14.4.3 Linkage Disequilibrium MAS (LD-MAS)

LD-MAS is the 'next-best' type of MAS, and performed when a particular marker allele or multiple marker haplotype is found to be in population-wide disequilibrium with the functional mutation at a QTL. In practical terms, the biggest difference to GAS will be the fact that a number of markers will need to be genotyped (LD haplotype) rather than a single marker. Over the long-term however, it is very important to monitor whether the association between the markers and mutation holds, or whether it is broken down by recombination. Like for GAS, the effect of the haplotype should be confirmed in multiple populations or selection lines, to ensure it is universally applicable. Different effects may be found for the LD markers as a result of different phase with the QTL allele and/or genetic background. A good example is a non-coding polymorphism in the ESR gene in pigs, which causes increased litter size in some breeds but has the opposite or no effect at all in others.

14.4.4 Linkage Equilibrium MAS (LE-MAS)

LE-MAS represent the form of MAS that is easiest to reach first (markers don't have to be so close to the QTL), but hardest to implement across the whole population. Since LE-MAS uses markers that are only in LD to a QTL within individual families, the phase between marker and QTL alleles must be re-assessed continuously and within each family in the MAS process. This requires genotyping for several markers of not only the selection candidates, but also many of their relatives to allow evaluation of the QTL effect. Although this implies that LD-MAS is dramatically more efficient, full-sib family sizes are very large for most aquaculture species, meaning that the re-estimation of marker-QTL association for every family is not so problematic as in livestock populations (Sonesson 2007).

14.4.5 Genomic Selection

A disadvantage of the LE-MAS, LD-MAS and GAS approaches is that only a limited proportion of the total genetic variance for a given trait is explained by the markers (QTL). An alternative strategy is to use a dense marker coverage across the entire genome to capture all the QTL, both large and small. This method is known as genomic selection, and is practically performed by dividing the entire genome up into chromosome segments, for example defined by adjacent markers, and then tracing all the chromosome segments (Meuwissen et al. 2001). A prerequisite for genomic selection is that the marker coverage is dense enough so that the effects of the chromosome segments will be the same across the population since the markers will be in LD with the QTL that they bracket. Previously limited to terrestrial livestock applications, this strategy is becoming relevant for aquatic species now that dense genetic maps containing tens of thousands of markers are becoming available (such as the Atlantic salmon 16,500 SNP chip developed by CIGENE in Norway). Implementation of genomic selection proceeds in two fundamental steps, 1. Estimation of the effects of chromosome segments in a reference population and 2. Prediction of genomic EBVs (GEBVs) for animals not in the reference population, for example selection candidates.

In a simulation of introgression schemes involving multi-trait selection comparing genomic selection with classical selection, genetic gain per generation for disease resistance may be doubled, and genetic gain for a production trait may also increase simultaneously (Ødegård et al. 2009). At the same time, the rate of inbreeding was reduced, with the largest effect in scenarios selecting on disease traits with low heritability. It was also concluded that for a given rate of inbreeding, higher selection intensities can be tolerated for genomic selection. The results indicated that genomic selection can provide a crossbred population with improved disease resistance combined with high productivity substantially faster than traditional selection schemes.

14.5 Other Applications of Genetic Markers

In addition to direct application of markers through MAS and genomic selection, they can also serve a number of other purposes connected to the direct management of breeding populations, and aquaculture production in the wider context. These are discussed in the following sections.

14.5.1 Parentage Assignment and Traceability

As discussed in Chapter 12, another potential use of highly polymorphic markers like microsatellites is the assignment of parentage to offspring, avoiding the need for physical tags. A related application of markers is the characterisation the relative contribution of broodstock in mass spawning events, introduced in Chapter 8. Selvamani et al. (2001), genotyped individual *Haliotis asinina* larvae by analysing five polymorphic microsatellite loci to identify the parents of individual larvae produced in three separate crosses. In all cases, the parents of an individual veliger could be determined from as few as three loci. The microsatellite analysis revealed that, in each of the crosses, a single male fathered most of the veligers, despite efforts to normalise the amount of sperm contributed by competing males. This highlights the inbreeding risks that some species face when certain breeding strategies are used without careful monitoring.

Traceability schemes for aquaculture species are of great importance for tracing market product to farm of origin in the event of detection of disease or toxins in fish or shellfish in the marketplace. DNA markers can be used to trace such events by sampling and genotyping live individuals or products at any stage along the production chain. The most suitable and cost effective traceability strategy for a particular industry will depend heavily on the organisation of that industry, for example the degree of recording transfer of fish, eggs and larvae between tiers. Based on simulation studies, Hayes et al. (2005) concluded that even if complicated logistics prevent the adoption of marker based schemes by some industries, traceability with DNA markers may still be important for verification of labelling-based schemes.

14.5.2 Genetic Interactions

Considerable public concern exists about the impact that escaped farmed fish have on wild populations. Genetic interactions, where escaped farmed fish potentially interbreed with wild fish, are not well understood at present in terms of their scope and consequences. Whilst not directly impacting selective breeding schemes, the potential for displacement of wild populations due to the invasion of escapees would remove a natural source of genetic variation that may be desired to be accessed again in the future. Genetic markers provide a useful tool to investigate the extent of the impact escaped farmed fish have on wild populations, though identification of escapees and potential interbreeding. In a study of wild and domesticated salmon in Norway, 12 microsatellite markers were used to investigate the discrimination of wild and farmed fish (Skaala et al. 2004). Assignment tests indicated that the wild and domesticated salmon could be distinguished with high precision. Less than 4% of domesticated salmon were misassigned as wild salmon, and less than 3% of wild fish were misassigned as domesticated salmon. Fish from individual domesticated strains were identified with similarly high precision. Assignment to wild salmon stocks was less accurate, with the exception of the sample taken from the river Neiden, where 93% of the individuals were correctly assigned.

14.5.3 Genetic Variation

Genetic markers provide a useful means of monitoring levels of genetic variation in farmed populations (Notter 1999), especially in situations where highly structured and controlled breeding programs are not implemented. In some cases, such studies have revealed large declines in genetic diversity in culture when compared to wild ancestral populations (Smith and Conroy 1992). Evans et al. (2004) used microsatellite markers to investigate levels of genetic diversity within cultured populations of *Haliotis midae* and *Haliotis rubra* in South Africa and Australia. The observed loss of alleles in both farm samples was significantly greater than that expected due to genetic drift based on such an effective population size, and highlighted the need for genetic monitoring of aquaculture hatchery systems to ensure that sufficient numbers of known pedigree broodstock are utilised in each generation. Indeed, results from this type of analysis may highlight the need for structured and well-planned breeding programs to farmers.

14.6 Gene Expression Data

Apart from the use of gene expression profiles to better understand the function of genes underlying important commercial traits, Robinson et al. (2008) have proposed the use of gene expression profiles as an indirect test for disease resistance. The concept is that certain cells (e.g. macrophages or leukocytes) will respond to the disease agent, and the resulting cascades of gene expression changes in these cells could potentially differ between animals that are able to resist versus those that are more susceptible to the disease. Such an approach could use high-throughput measurement technology like microarrays to identify likely indicator genes using a modest calibration data set consisting of samples from extreme performing animals for the disease challenge. After identification of candidate genes, expression values from a larger set of animals could be measured using more suitable platforms. As tissue for gene expression profiling can potentially be collected from live candidates, the profiles could then be used as a selection tool. The use as a selection tool assumes that a method of eliciting a gene expression response from breeding candidates is available, for example by challenging cells derived from these individuals to disease.

Robinson et al. (2008) evaluate a method for formulating prediction equations using random regression with cross validation on a set of gene expression data from breast cancer patients, and found a moderate correlation between predicted and actual phenotype (0.32 ± 0.06). Based on simulations using gene expression data in a selective breeding program, Robinson and Hayes (2008) found that disease resistance was doubled after six to seven generations of selection, and varying the phenotypic and genetic correlation had a relatively small effect on the overall genetic response after 10 generations. Benefit–cost was positive under all scenarios. With 10 generations of selection under the optimal scheme, the model predicted a benefit–cost ratio of more than 17:1.

14.7 Transgenics

Recent advances in technology have facilitated the artificial manipulation of genes and chromosomes in living organisms. Naturally, the prospect of transgenic fish and shellfish that perform dramatically better for important commercial traits is of great interest to both researchers and the industry (Zbikowska 2003; Dunham 2004). Some of the major areas of focus for transgenic research in fish have been the use of growth hormones (GHs) to increase growth and feed conversion efficiency, antifreeze proteins (AFPs) for enhanced cold tolerance and freeze resistance, antimicrobial peptides for increased disease resistance, metabolic genes to promote alternative diets, and genetic methods for inducing sterility. Research into transgenic methods in fish has been performed for many years, and the methods have been well-developed. Mollusc and crustacean species however, pose their own particular challenges surrounding the introduction and expression of foreign genes. Some common methods used to produce transgenic animals include:

- Micro-injection of DNA into a newly fertilised zygote
- Transfer of a nucleus from a somatic cell into an enucleated oocyte
- Sperm mediated gene transfer
- Electroporation and intra-testicular injection of transgene DNA
- Retroviral-mediated transfer of transgenes.

Some particularly notable results from transgenic studies in fish include that of Fletcher et al. (2004), who reported the successful expression and inheritance of the GH gene through six generations of Atlantic salmon. In a study of transgenic Nile tilapia (*Oreochromis niloticus*) that over-expressed the GH gene throughout their bodies, Kobayashi et al. (2007) found that the food-conversion efficiency of the transgenic fish was 35% higher than that of their non-transgenic siblings, the rearing period required for the transgenic fish to reach a body weight of 20 g was about 75% of that required for non-transgenic fish that were fed the same type and quantity of food, and the total amount of ammonium-nitrogen excreted by the transgenic fish was about 69% of that excreted by the wild-type fish over their lifetime. The transgenic salmon reach market size approximately one year earlier than non-transgenic farmed salmon. However, the effect of introducing a growth-hormone gene construct into fish to increase growth rates appears to be dependent on the degree to which earlier enhancement has been achieved by traditional genetic selection. This has been documented by Devlin et al. (2001), who found that the growth of transgenic wild strain rainbow trout did not surpass that of a fast-growing non-transgenic domesticated strain of trout used in aquaculture. The results indicate that similar alterations of growth rate can be achieved both by selection and by transgenesis in rainbow trout, but that the effects are not always additive.

Other species have had similar encouraging results with transgenics. Rahman and Maclean (1999) produced three lines of transgenic tilapia harbouring a novel piscine growth hormone (GH) gene construct containing a chinook salmon growth hormone gene spliced to ocean pout antifreeze gene regulatory sequence. Expression of chinook salmon GH was demonstrated in G0, G1 and G2 transgenic fish in these lines and resulted in dramatic growth enhancement. The average weight of the G1 and G2 transgenic fish was found to be three times ($P < 0.001$) greater than that of their non-transgenic siblings.

Despite these successes, in the short term, it is unlikely that transgenic animals will play a major part in aquaculture production, however in the longer term, they are likely to be important. Transgenic animals are a potential way to provide rapid introduction of new phenotypes/genotypes into elite animals, and the generation of novel phenotypes of substantial value. Standing in the way of more widespread introduction is the major issue of consumer acceptance of transgenic animals. In many plant industries, transgenic plants have faced substantial opposition because of public concerns based on environmental impact. There is no doubt that advances in genomics, chromosome manipulations and gene expression technology will mean that larger scale production of transgenic individuals will be possible in the future. However, the questions regarding consumer acceptance and risk to natural populations may take far longer to resolve, limiting the practical implementation of such technology. Given that some results show that traditional selection can achieve the same result in some cases, it is likely that this will be the preferred method in industry for some time to come.

14.8 Genome Sequencing and Future Technologies

Dramatic advances in DNA sequencing technology have meant that the cost and time required to perform whole-genome sequencing has been reduced by several orders of magnitude. At the time of writing, three major platforms represent the forefront of DNA sequencing technology (Roche 454™, Illumina Genome Analyzer™ and ABI SOLiD™), each offering hundreds of times greater throughput and reduced cost than traditional Sanger sequencing. As a result, genome sequencing initiatives that were previously limited to humans and model species, are now underway for many livestock and a number of aquaculture species. At the time of writing, such projects are underway for Atlantic salmon and Atlantic cod, and are in the planning stages for a number of other species. DNA sequencing technology continues to advance rapidly, and it is likely that whole-genome sequencing could soon become a routine activity for even small scale labs equipped with a single ultra-high throughput sequencer. Having such genomic information available for aquaculture species will be an extremely valuable resource, providing dramatically increased possibilities for identifying important genes underlying commercial traits and characterising the polymorphisms responsible for variation in these traits. Ultimately, for selective breeding programs, this means that far more powerful tools will be available to increase genetic gain and manage populations to ensure long term sustainability and productivity.

Chapter 15
Reproduction Techniques

15.1 Introduction

In addition to the quantitative genetic approach where genetic variation is exploited and used to produce genetic gain each generation, there are possibilities to make more fundamental changes to the animals chromosomal arrangements by manipulations during reproduction. In most aquatic species, fertilization takes place externally and it is possible to manipulate the number of chromosomes in the progeny to produce haploids, triploids and tetraploids.

Key to the process of chromosomal manipulations is the development of eggs and process of meiosis, as illustrated in Fig. 4.1. During meiosis, the genome of a diploid germ cell, which is composed of long segments of DNA packaged into chromosomes, undergoes DNA replication followed by two rounds of division, resulting in four haploid cells. The development of the eggs starts with a diploid cell known as an oocyte. In meiosis I, chromosome doubling occurs, forming sister chromatids, held together by centromeres. The chromosome copies are than paired, so that the homologous chromosomes are aligned side by side and total four in number. Subsequently, the homologous chromosomes are separated as the cell divides in two, forming two cells with two sets of chromosomes each. During meiosis I, recombination between the homologous chromosomes takes place. Only one of the cells develops further into a secondary oocyte, and the other cell, known as the first polar body, is lost. Meiosis II is completed after fertilisation, when the secondary oocyte divides. The second polar body is then lost while the fertilized egg develops into a diploid individual.

15.2 Gynogenesis

Gynogenesis is a special form of reproduction that has been known to occur in nature. During gynogenesis, the egg is activated by a genetically inert sperm that has been treated with ionising or ultraviolet radiation. At this stage, the second polar body is still inside the egg, waiting to be shed. Treatment of the egg by an

T. Gjedrem, M. Baranski, *Selective Breeding in Aquaculture: An Introduction*, Reviews: Methods and Technologies in Fish Biology and Fisheries 10, DOI 10.1007/978-90-481-2773-3_15,

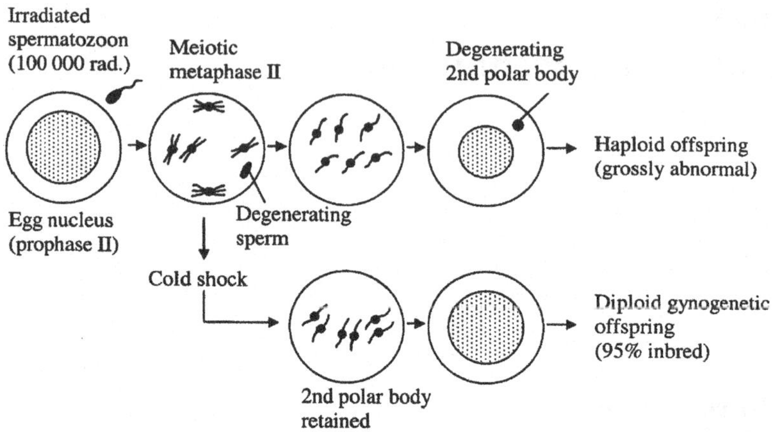

Fig. 15.1 Schematic presentation of induced gynogenesis. Reproduced from Purdom (1993) by permission of Chapman and Hall

environmental shock such as cold, heat, pressure or chemical shock, results in the second polar body being retained in the egg, forming a diploid egg. The two sets of chromosomes are close to identical, having originated from the same egg cell. The difference between the two sets of chromosomes is solely the result of the recombination phase in meiosis I (Fig. 15.1). However, this means that gynogenetic animals are highly inbred. As a result of crossing-over, the inbreeding will not be complete and Cherfas (1981) estimated the inbreeding coefficient to be $F = 0.60$ after one generation of gynogenesis. In Fig. 15.1 Purdom (1993) give a $F = 0.95$ which is unlikely. However, if gynogenesis is repeated for several generations the animals will soon be completely inbred ($F = 1.00$).

Cold shock was the first treatment used to induce diploid gynogenesis (Purdom 1972), and was performed by holding newly fertilised eggs in a mixture of water and ice. The application of cold shock produced varying results in salmonids, leading to the adoption of heat shock as an effective alternative (Chourrout 1980). A third method, hydrostatic pressure, is now commonly used to induce diploid gynogenesis in fish.

To produce a high percentage of diploid gynogenetic fish, the timing of shock is very important. The shock is usually started 5–15 minutes after fertilization with the inert sperm, and lasts for up to 20 minutes using heat shock or between 20 and 120 minutes using cold shock. Refstie (1983b) found that the highest frequency of gynogenetic fry was produced when eggs of Atlantic salmon and rainbow trout were heat shocked at 24 and 26°C, respectively.

Since the gynogenetic animals are highly inbred, they suffer from inbreeding depression that is particularly manifest as poor survival and low growth rate. In species where females are the homogametic sex, like salmonids, all–females are produced with only genetic information from their dams.

Gynogenetic animals may be of interest in breeding programs for producing highly inbred lines that can be subsequently crossbred to exploit hybrid vigour. The highly inbred nature of gynogenetic animals means that they show very low genetic variation, and may therefore be of particular interest for laboratory experiments (Komen 1990).

15.3 Androgenesis

In the process of gynogenesis, the sperm is inactivated, while in andrognesis the chromosomes in the egg are inactivated. However, applying ionising radiation to eggs may be problematic since it can result in damage to the eggs' contents and therefore can result in reduced egg viability. To obtain successful embryonic development, the treated eggs are then fertilised by normal sperm. Thorgaard et al. (1990) produced androgenetic rainbow trout by suppression of the first cleavage of eggs followed by sperm irradiation and fertilisation from tetraploid males. Androgenic animals are of interest for breeders and researchers for similar reasons as gynogenetic individuals.

15.4 Triploidy

To induce gynogenesis, a shock is applied to block secretion of the second polar body. If this is performed on eggs fertilised by viable sperm, the second polar body will be retained and a triploid animal will be produced with two sets of chromosomes from the egg and one set of chromosomes from the sperm (Fig. 15.2). Such triploids are usually sterile. In salmonids, this is typically observed as males producing milt that is not fertile, while females lack gonad production.

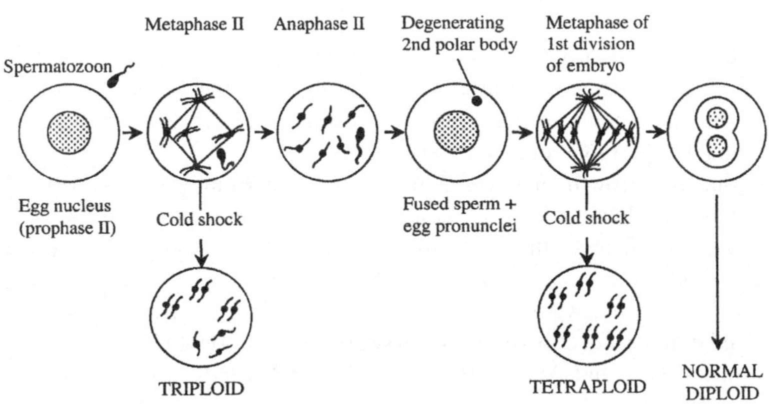

Fig. 15.2 Diagrammatic representation of induced polyploidy. Reproduced from Purdom (1993) by permission of Chapman and Hall

Triploid fish and shellfish are used for farming in some countries for the following reasons:

- Since the animals are sterile, escapees will not breed with and affect wild stocks
- In most species with early sexual maturation, triploids are of interest since they are sterile and will continue to grow while the diploids will produce gonads.

The effects of triploidy vary from trial to trial and between species. Generally, survival of triploids during hatchery period is poorer than diploids. The advantages of triploids are most obvious after the diploids start to produce gonads. However, triploids do not appear to be universally superior across all species. For growth rate, there are a number of reports showing no or small effects of triploidy, including tilapia (Pechsiri and Yakupitiyage 2005; Hussain et al. 1995), Atlantic salmon (Oppedal et al. 2003; Friars et al. 2001) and rainbow trout (Jonasson 1984).

In shrimp (*P. chinensis*) Xiang et al. (2006) report that triploids did not show higher growth during the immature stage, but exhibited superior growth during the maturation stage. Characteristics of the reproductive organs indicated that triploid shrimp may be sterile and that the sex ratio can be changed through triploidisation.

In shellfish, it is generally accepted that triploids perform similarly to diploids before the beginning of maturation (Guo and Allen 1994a). This was true for the catarina scallop, where the growth advantage was first seen after 146 days. At 382 days of growth, the average total weight of triploids was 18% larger than diploid controls and they had 37% greater muscle weight (Ruiz-Verdugo et al. 2000). Similar results were reported for Pacific oysters (Guo et al. 1996), where diploids and tetraploids were crossed and had comparable survival to diploids but were 13–51% larger than their diploid counterparts.

Hand et al. (2004) compared the effect of triploidy in oysters selected for fast growth over three generations and in a control line. The experimental groups were tested in three different environments, Cromarty Bay, Tea Gardens and Tilligerry Creek in Port Stephens, Australia. Growth curves of the different groups from Cromarty Bay are shown in Fig. 15.3. The results from the other two groups, Tea Gardens and Tilligerry Creek, were very similar.

Averaging the growth of oysters in the three environments, the triploid control line was 36% heavier than its control line, and the triploid selected line (L2 triploid) was 44% heavier than the diploid selected line (L2 diploid). On average, the triploid selected line (L2) was 74% heavier than the control diploid line, indicating that growth improvements from selective breeding and triploidy were at least additive and could reduce the time to market by at least 10 months.

In molluscs, Guo and Allen (1994a) concluded that triploids were significantly larger than diploids in almost all species studied. In mollusc, triploids are of particular interest due to their reduced gonadal development which prevents deterioration of meat quality.

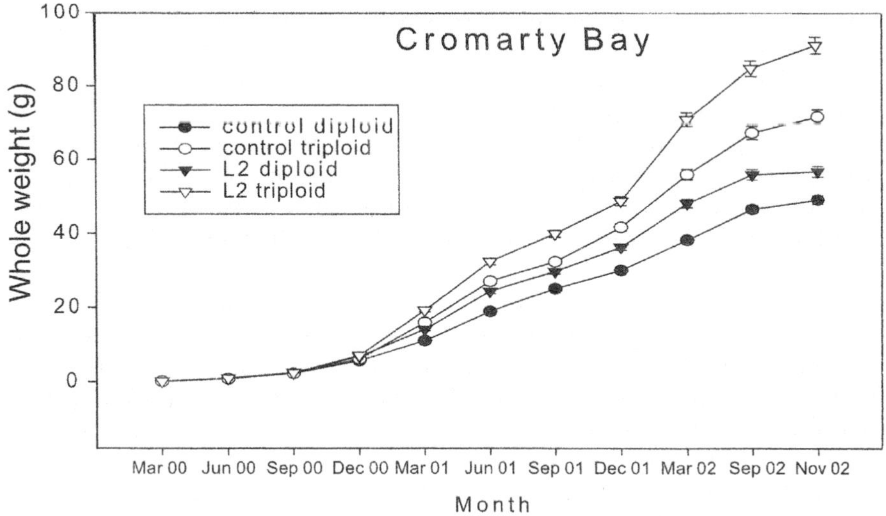

Fig. 15.3 Whole weights of diploid and triploid control and selection lines of Sydney rock oysters, *Saccosstrea glomerata*, at Cromarty Bay in Port Stephens, Australia from March 2000 – November 2002. Reproduced from Hand et al. (2004) by permission of Elsevier Science

15.5 Tetraploidy

A tetraploid animal has four sets of chromosomes instead of the diploid animals two. Tetraploids are of primary interest for the subsequent production of triploids, through crosses with diploids. Successful production of tetraploids in rainbow trout was obtained by the suppression of first cleavage (Fig. 15.2). Chourrot (1984) and Myers et al. (1986) were some of the first to successfully produce tetraploids. In 1993; Guo and Allen produced tetraploids of Pacific oysters by inhibiting the first polar body of eggs from triploids that had been fertilised with sperm from diploids (Guo and Allen 1994b). Eudeline et al. (2000) refined the method developed by Guo and Allen by varying the duration of the treatment to inhibit polar body I of triploid eggs. This produced clear effects on the ploidy of the progeny.

15.6 Production of Single Sex, YY and XX Stocks

In most aquatic species, males grow faster than females. An exception to this rule is the Atlantic halibut where females grow faster. Therefore, single sex production can be advantageous for production fish. The greatest focus on single sex production in aquatic farmed species is in tilapia, since they become sexually mature very early and reproduce in grow-out ponds, causing overpopulation and stunting. Several methods have been investigated to reduce this problem and produce monosex progeny in this species:

- The addition of male hormone (17 α-methyl-testosterone) in first feeding produces close to 100% phenotypic males
- The addition of female hormone (diethylstilboestrol) in first feeding produces close to 100% phenotypic females
- The crossing of certain tilapia species yields a higher frequency of males, for example *Oreochromis niloticus* x *O. aureus* and *O. niloticus* x *O. hornorum*
- Trained personnel can sex tilapia fingerlings and thus separate males and females
- The production of monosex populations with YY or XX chromosomes.

There are several ways to produce genetically monosex broodstock. Mair et al. (1997) describe a method applied in tilapia where genetic males were feminised to phenotypic females by the administration of β-estradiol in first feeding period. Eggs from the feminized genetic males (XY) were fertilized with sperm from normal males (XY) which produced the following progeny groups; 25% females (XX), 50% normal males (XY) and 25% males (YY). Identification of YY males can be performed through progeny testing. When the YY males are identified, eggs from ordinary females (XX) may be mated with sperm from YY males to obtain only male progeny. In practice, 95% males have been obtained using this method, and the growth performance of such all-male populations increased by up to 58% (Mair et al. 1995).

Chapter 16
Economic Benefits of Breeding Programs

16.1 Introduction

The primary aim of a breeding program is to produce biologically efficient animals that grow faster, have higher survival and improved product quality. Animals that have higher growth rate (Table 16.1) will need less feed per kg of body weight and therefore more efficiently utilise feed resources. In freshwater aquaculture, this leads to more efficient use of both land area and water resources, a particularly important result given the limited nature of such resources in many areas. As higher turnover is made possible by faster growth of the animals, production levels within the available facilities will increase. All these improvements in efficiency add up to increased profitability for the farmers, and increased availability and cheaper prices for the consumer.

However, the ultimate efficiency of production, both in terms of yield and economics, depends on many factors, not just gains resulting from genetic improvement. Such factors include optimal environmental conditions, balanced feeds meeting the animals' nutritional requirements, good management and husbandry practices, and willingness in the market to pay for the product. With so many factors influencing the cost of production, it is not easy to separate the contribution of each factor, and little has been published on this subject in aquaculture species. This chapter discusses the added value and economic benefits of running breeding programs and the economic value of the genetic gain obtained.

16.2 Cost of Broodstock Production

In meat producing terrestrial livestock species as well as for carnivorous fish species, the cost of feed usually represents more than 50% of the total production cost. There are however, large differences between species in the cost of feed necessary to produce and maintain broodstock. In Table 16.2, some estimates of the relative cost of feed consumed by broodstock are given. For sheep, the feed consumed by female breeders represents approximately 70% of total feed costs, with figures of 55% for

T. Gjedrem, M. Baranski, *Selective Breeding in Aquaculture: An Introduction*, Reviews: Methods and Technologies in Fish Biology and Fisheries 10, DOI 10.1007/978-90-481-2773-3_16,

Table 16.1 Production time in months of Atlantic salmon to reach body weight of 4 kg

	During 1970s	During 2000s
Freshwater	16	8
Seawater	24	12
Total	40	20

Table 16.2 Relative cost of feed consumed by female broodstock for different species, expressed as a percentage of total feed costs

Beef cattle	Sheep	Pigs	Poultry	Salmonids	Reference
52	72	33	10		Large (1976)
58	67	36	12		Dickerson (1978)
				1–5	Kinghorn (1983)

beef cattle, 35% for pigs and 11% for poultry. In salmonids, which are carnivorous, Kinghorn (1983) estimated that the amount of feed consumed by broodstock to be as low as 1–5%. The main reason this estimate is so much lower than other livestock species is the fact that salmonids are highly fecund, and far fewer broodstock are required.

In terrestrial livestock, selecting for high growth rate at market size is problematic since it creates a correlated response in sexually mature weight. Larger animals naturally demand more maintenance feed, however in aquatic species this is not a major problem since the cost of feed for broodstock is relatively low due to the species' high fecundity.

16.3 Cost of Running a Breeding Program

The cost of running a breeding program is difficult to estimate because of the inherent differences between particular situations both within and across species. These variables include differences in program dimension, number of traits included in the breeding goal, and testing procedures. A breeding program that uses sib information when selecting breeders needs to test a large number of families in each generation. When physical tags are used to identify the animals, it is necessary to invest in large numbers of tanks or hapas to keep families separate until the juveniles reach a body size suitable for tagging. These investments could be reduced by immediately placing offspring in a communal environment and assigning parentage with DNA markers, rather than physical tags. However, DNA testing also involves costs, and some of the breeding candidates need to be physically tagged so they can be traced easily at mating.

Kontali Analyse (2004) estimated the distribution of the total economic value of the Norwegian production of Atlantic salmon in 2004 (564,000 tons) (Table 16.3).

Table 16.3 Estimated distribution of the total economic value of the Norwegian production of Atlantic salmon in 2004

Component	Economic value (million NOK)
Total value	9,710
Nucleus breeding	40
Egg production	70
Smolt production	1,200
Grow-out production	8,400

These figures highlight that only 0.41% of the total value or 0.07 NOK per kg of fish produced, was returned to the breeding companies running the breeding program where the genetic gain was produced. This may be taken as an estimate of the cost of running the breeding programs in Norway and should be reduced by the profit of the company.

16.4 Economic Benefit of Breeding Programs

Relatively few studies have been performed to assess the economic value of breeding programs in aquaculture. Gjerde and Olsen (1990) studied the economic value of running a breeding program in Atlantic salmon. They estimated the genetic gain to be 10% for growth rate and 3% for age at sexual maturation, corresponding to a profit of \$0.13 and \$0.08 per kg fish respectively, totalling to a \$0.21 per kg fish improvement overall.

Gjerde et al. (2007b) estimated the economic benefit of Norwegian breeding programs for Atlantic salmon. Close to 100% of Atlantic salmon production in Norway is based on improved stocks, and selection has been performed over six to eight generations (Table 2.1). In most recent generations, this selection has been performed for growth rate, disease resistance and product quality, in total seven to eight traits in each program. However, estimates of genetic gain are only available for growth rate and to some extent for feed conversion efficiency and early sexual maturation, as discussed in Chapter 3.

For feed conversion efficiency, Gjerde et al. (2007b) used a simple calculation assuming that the current generation selectively bred salmon use 25% less feed per kg of body weight gain than the offspring of wild salmon (Thodesen et al. 1999), and a feed cost of 8 NOK per kg. The accumulated economic value of increased feed efficiency accounted for 3 NOK per kg of fish produced. The economic value of the improved growth rate is the reduced cost of production due to shorter production time to harvest, which was halved from 40 to 20 months over seven to eight generations of selection (Fig. 16.2). Additionally, further economic benefits arise from the lower proportion of early sexually maturing fish, reduced production losses due to increased disease resistance, and improvements in fillet colour and fillet fat yield. The total economic value of the selective breeding work for Atlantic salmon in Norway over seven to eight generations of selection was estimated to be at least 15 NOK per kg of fish produced, equating to around 2 NOK per generation, or 0.5 NOK per year. Considering the total production of Atlantic salmon in Norway in 2007 was 736,000 tons, this represents genetic gain with an economic value in the order of 368 million NOK per year (at present 7 NOK approximately equals 1 US$).

If the running costs of the breeding program for Atlantic salmon are around 0.07 NOK per kilogram of fish produced (according to Kontali Analyse (2004)), the benefit/cost ratio is 0.50/ 0.07, or 7–1. This figure is substantially lower than an earlier estimate of benefit/cost ratio of 15–1 by Gjedrem (1997).

Ponzoni et al. (2007) investigated the benefit/cost ratio of a breeding program for Nile tilapia over a period of 10 years or 10 generations. The effect of different factors on the benefit/cost ratio was investigated. It was found that the heritability value of the traits under selection had a moderate effect, the level of investment and annual cost of operation had a relatively small effect while the effect of the market price of fish was substantial. The greatest contribution to the benefit/cost ratio was reproductive efficiency. The lowest benefit/cost ratio, 8.5 to 1, was obtained when each fish produced 10,500 progeny per year and the highest benefit/cost ratio, 60 to 1, was obtained when the fertility was 130,000 offspring per female through repeated spawning. This demonstrated that even under the most conservative assumptions, genetic improvement programs are highly beneficial from an economic point of view.

In terrestrial livestock species, a range of different benefit/cost ratios have been estimated. Full investment appraisals for sheep, pigs and cattle show net benefit/cost ratios ranging from 5 to 1 and 50 to 1 (Barlow 1983; Mitchell et al. 1982). These estimates clearly show that investment in selection programs for terrestrial livestock yields high returns, expressed in terms of interest on invested capital.

Despite the fact that benefit/cost ratios appear to be universally high in both livestock and aquatic species, the large variation both between and within studies (from 5/1 to 60/1) shows that it is a difficult task to obtain unbiased estimates. One possible reason for this variability is that the data used for estimation may have been unsuitable for the purpose. In any case, real differences in benefit/cost ratio are to be expected as the result of differences in the design of breeding programs and the efficiency with which they are run. Given that generation interval plays a major role in the rate of genetic improvement, it is not surprising that this factor has a large influence on benefit/cost ratio.

16.5 Relative Contribution of Selection and Feed Regimes to Performance

For fish and shellfish species, there are a few good estimates of the contribution of selective breeding to the obtained improvement relative to the cost of feed and other production costs. A Norwegian committee discussing possibilities for value creation in aquaculture concluded that selective breeding had been the most important single factor for the development of the industry (Jensen et al. 1999). A group of managers

Table 16.4 Growth rate of broiler strains from 1957 and 2001 fed typical diets from 1957 and 1991 by period of 84 days

Strain from	Typical feed from 1957	Typical feed from 1991
1957	1.43	1.61
1991	4.48	5.52

Reproduced from Havenstein et al. (2003) by permission of Poultry Science Association

in the aquaculture industry stated that the three main sources for development in salmon farming were feed, technology and selective breeding, each with an approximately equal contribution (Norsk fiskerinæring, 8:1995).

In poultry, Havenstein et al. (2003) studied the relative performance of broiler strains from 1957 and 2001 fed on typical diets from the same years (Table 16.4). Growth over 84 days increased from 1.43 to 5.52 kg, representing a 3.9 fold improvement.

Comparing the relative contribution of each of these two factors to this improvement, Havenstein et al. (2003) concluded that genetic selection was responsible for 85–90% of the growth rate improvements in broilers over the past 45 years, while nutrition has been responsible for 10–15% of the improvement.

Feed conversion ratio (FCR) was also substantially improved over this period, with a reduction from 3.84 to 2.72 kg feed/kg growth. Within this overall reduction of 29%, the contribution from selective breeding was approximately equal to the contribution from improved feed.

Over the 84 day trial, mortality was nearly twice as high in the modern strain compared to the 1957 strain (Havenstein et al. 2003), however there was no difference in mortality of birds of the same size.

16.6 Who Benefits from Genetic Improvements?

16.6.1 The Animal

As discussed earlier, the process of domestication commences immediately in the first generation of captivity. One of the most obvious effects of this domestication is the reduction in stress and fearfulness that the animals exhibit. In the freshwater stage of their life-cycle, salmonids tend to form hierarchy systems where some fish are dominant. This is apparent from the right-skewed non-normal distribution of body weight of fingerlings (Fig. 16.1). The distribution of Atlantic salmon was most extreme of the species investigated.

This skewed distribution in Atlantic salmon fingerlings is illustrated by means of the coefficient of variation (CV) in Table 16.5. The CV was found to be most extreme in the zero generation where wild broodstock were used as parents, and was observed in all four populations studied. The CV of body weight varied from 75 to

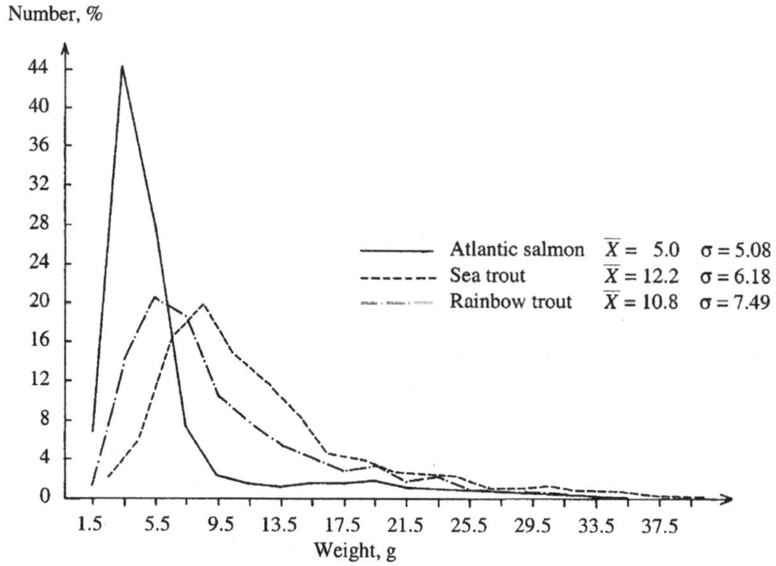

Fig. 16.1 Body weight distribution of fingerlings of different salmonid species (Gjedrem 2005)

Table 16.5. Average body weight (x) and coefficient of variation (CV) in four populations of Atlantic salmon parr for four generations of selection for body weight after two years in sea cages

		Population							
		1		2		3		4	
Generation of selection	Year-class	$\overline{X}$	CV	$\overline{X}$	CV	$\overline{X}$	CV	$\overline{X}$	CV
0	72–75	6.1	78	17.7	75	5.5	84	7.8	75
1	76–79	8.8	67	3.7	55	4.3	74	5.5	59
2	80–83	4.8	59	6.8	48	6.4	58	6.3	64
3	84–87	6.4	50	5.2	40	5.7	51	8.2	56
4	88–91	5.6	43	5.8	47	8.3	43	12.5	42

Reproduced from Gjedrem and Fjalestad (1997) by permission of AKVAFORSK.

84% in this generation. On average, the CV was reduced in each generation of selection. After four generations of selection, the distribution of fingerling body weight approached normality and the average CV of the four populations was reduced to 44%. These results indicate that the domestication process tends to reduce aggression levels in some fish, breaks down the hierarchy systems, and produces a more uniform product in both growth and behaviour.

Moberg (2000) discussed the implications of stress on animal welfare and stated: ‘Gradually we have come to accept that animals also suffer from the burden of stress, and that when suffering from stress they develop very similar pathologies.

Like humans, while experiencing severe stress, animals can succumb to disease or fail to reproduce or develop properly'.

Given that stress can negatively impact welfare and productivity, methods to measure and quantify stress are of interest to breeders. The best indicator and easiest trait to measure is considered to be cortisol levels in the blood. In a large data set of Atlantic salmon and rainbow trout, Fevolden et al. (1993) estimated genetic variation in cortisol levels following standardised confinement stress. Low heritability for cortisol level was found for Atlantic salmon ($h^2 = 0.05$) and low to medium in rainbow trout (mean $h^2 = 0.27$). Fevolden et al. (2002), subsequently estimated the realised heritability for cortisol levels to be $h^2 = 0.50$ in rainbow trout. An interesting result was that the cortisol levels in Atlantic salmon were on average twice as high as in rainbow trout. However, baseline cortisol levels in wild salmon and trout are not known. The difference observed between the two species could be related to the fact that rainbow trout have been farmed for around 30 generations in Norway and have reached a higher level of domestication than the Atlantic salmon, which had been farmed for only six generations when the investigation took place.

Fevolden et al. (2002) selected for high and low post-stress levels of plasma cortisol and tested the correlated response to growth rate in rainbow trout. Superior growth performance was found in the low cortisol line compared to the high responding line. Fevolden et al. (1993) challenged similar high and low selection lines of Atlantic salmon with three bacterial pathogens; *Aeromonas salmonicida* which causes furunculosis, *Vibrio samonicida* causing cold-water vibriosis and bacterial kidney disease (BKD). Mortality due to furunculosis and vibriosis showed opposite trends. There was a higher mortality following the vibriosis challenge in the high-stress line, but lower mortality in the same line following the vibriosis challenge. For the BKD challenge, mortality levels were similar in both lines.

16.6.2 The Farmer

Aquaculture production is commonly divided into two sectors; the fingerling producers and the grow-out sector. Large integrated companies cover the whole production cycle. In the hatchery sector, the genetically improved stock will generally show higher survival and growth until the fingerling stage. In Atlantic salmon, a survival level of 30% from eyed egg to smolt stage was considered acceptable in the first one to two generations of production, while today it is common to obtain 80–90% survival (Håvard Bakke pers. comm.). This dramatic improvement is most likely the result of the domestication and selection process together with better feed and management.

Increased growth rate implies that the animals will reach market size in a shorter time. Faster growth and shorter production time will reduce feed costs, but since feed costs account for a relatively small part of total production cost in fingerlings, around 15% (Torbjørn Åsgård pers. comm.), genetic improvement of feed utilisation is of relatively small benefit to smolt/fingerling producers.

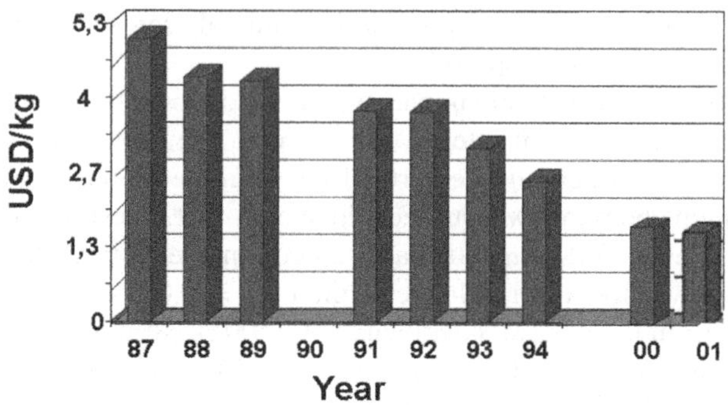

Fig. 16.2 Production cost of Atlantic salmon in Norway (Torbjørn Åsgård pers. comm.)

In the grow-out sector there has been a marked reduction in production time (Table 16.1) and improved feed conversion ratio, which has reduced the cost of production dramatically. Figure 16.2 highlights the dramatic reduction in the production cost of Atlantic salmon in Norway over a 24 year period. The main part of this reduction has taken place in the grow-out sector. This dramatic improvement in production efficiency is a result of many factors, however it is apparent that the single most important factor has been the genetic improvement of the fish. According to Gjerde et al. (2007b), it is evident that most of the economic benefit of the selective breeding work is harvested by the grow-out producers and consumers, and not by the breeding companies developing the genetically improved stocks.

As discussed earlier, more than half of the aquaculture production in the world takes place in freshwater, predominantly in earthen ponds. Production of shrimp, tilapia and carp species for example, frequently use land area where rice or other cereal production is an alternative. As growth rate and survival of animals are improved, the time to reach market size is reduced (Table 16.1). Therefore, the overall turnover rate increases and a given unit of land area has a higher relative production. This is important given that simply expanding the land area used for production is not a viable alternative in many areas where land and water resources are limited.

Increased productivity is of particular importance in areas with a shortage of clean water, and as an added benefit, increased production using less fresh water reduces the amount of effluent from farms that requires treatment. A reduction in water usage can also reduce the overall quantity of 'polluting' discharge from farms.

The increased productivity arising from genetic improvement automatically reduces the cost of production (Fig. 16.2). For consumers, the most obvious result is a reduction in the price of aquaculture products. This has been particularly apparent in Atlantic salmon, which made the transition from a high-priced luxury product to a cheap and plentiful commodity as a result of production improvements.

In conclusion, the whole value chain benefits from the genetic improvement in the form of increased predictability and reduced use of resources. However, given the variability that occurs over time, the particular members of the value chain that benefit most from the improvement in productivity, such as the fingerling producers, grow-out farmers, consumers and the wider community, tends to vary over time.

16.7 Ownership of Genetically Improved Material

It is obvious that breeding companies need some form of legal or biological protection to assure revenues from genetic improvement and return on their investments related to development of genetically improved stocks. However, this issue is complicated given the many levels within the production chain that genetically improved material passes through. The Research Council of Norway discussed this matter in a research program report:

'Different biological protection measures related to access to and exclusive rights of improved genetic material have been discussed (Rosendal et al. 2005). Alternative biological protection mechanisms can be:

(a) continuous upgrading and documentation of the genetic material so that breeding company is always ahead of its competitors.
(b) crossbreeding and sale of hybrids; or sale of sterile production animals.

Legal protection can be secured through:

(a) branding and trademarks.
(b) material transfer agreements between the breeding company and the multiplier or farmer.
(c) patents.

Worth mentioning is that only single genes can be patented, not populations. It is worth noting that at present, patents can only be lodged on single genes and not populations, however this is an evolving area of discussion and debate.

An alternative approach is the establishment of a mandatory certification system based on the known pedigree of farmed fish (Rosendal et al. 2005). This could be an integrated part of a larger traceability system for farmed fish production (Slettan 2004). The verification of such a system could be performed using different types of genetic markers (Haugen Rengmark et al. 2006; Hayes et al. 2005; Skaala et al. 2004; Villanueva et al. 2002) and at a relatively low cost because it would only require the genotyping of alleged parental fish. Such a system would allow the breeding companies to check the likelihood that fish originated from their population. A balanced strategy combining such traceability tools, together with branding and trademarking, continuous upgrading of the genetic material through selective breeding, and material transfer agreements, may be a viable option to address this issue. A dialog between breeding companies and public authorities is required to obtain legal approval of such a system'. (Reproduced from Gjerde et al. (2005) by permission of The Research Council of Norway).

Appendix

Portion selected (p), number of standard deviation from population average to point of truncation (x), selection intensity (i), for a normally distributed trait.

p%	x	i	p%	x	i
0.01	3.719	3.960	09	1.341	1.804
0.05	3.291	3.554	10	1.282	1.755
0.10	3.090	3.367	11	1.227	1.709
0.20	2.878	3.170	12	1.175	1.667
0.30	2.748	3.050	13	1.126	1.627
0.40	2.652	2.962	14	1.080	1.590
0.50	2.576	2.892	15	1.036	1.554
0.60	2.512	2.834	16	0.994	1.521
0.70	2.457	2.784	17	0.954	1.489
0.80	2.409	2.740	18	0.915	1.458
0.90	2.366	2.701	19	0.878	1.428
1.00	2.326	2.665	20	0.842	1.400
2.00	2.054	2.421	25	0.674	1.271
3.00	1.881	2.268	30	0.524	1.159
4.00	1.751	2.154	35	0.385	1.058
5.00	1.645	2.063	40	0.253	0.966
6.00	1.555	1.985	45	0.126	0.880
7.00	1.476	1.918	50	0.000	0.798
8.00	1.405	1.858	60	−0.25	0.644

Glossary

Additive effect The measured effect from individual genes that add together to give the total effect on a trait

Adenine A base found in DNA and RNA

All or none trait A trait that can be recorded as present or absent, (e.g. survival can be recorded as dead or alive)

Allele (allelomorph) One of several forms in which a gene may occur and are responsible for genetic variation

Allele frequency The frequency of occurrence of an allele in relation to that of other alleles of the same gene in a population.

Amino acid A molecule with the particularly important function of forming the basic building block making proteins, there are twenty of them

Androgenesis A special form of reproduction where progeny only inherit chromosomes (genes) from the male parent

Aquaculture Production of aquatic animals or plants under farming conditions

Artificial selection Selection of parents to reproduce for the next generation based on particular criteria by which we decide which animals shall be the parents that reproduce to make the next generation

Base population The individuals giving rise to a population (e.g. the individuals used for starting a breeding program)

Binary traits Traits that can be recorded in one of two alternative states, (e.g. sex as either male or female)

BKD Bacterial Kidney Disease

BLUP Best Linear Unbiased Prediction (e.g. of breeding values)

Breed A population of animals with a common genetic history and often some shared characteristics

Breeding The reproduction of animals or plants

Breeding goal A list of traits which shall be improved in a breeding program and their relative importance

Breeding program A plan for the selection and mating of animals to obtain a certain breeding goal

Breeding station The facility or unit where a breeding program is carried out

Breeding strategies Choice of basic approaches for selection (e.g. mass selection, combined selection) and mating (e.g. purebreeding, crossbreeding) in a breeding program

Breeding value The additive genetic performance of an individual according to the breeding goal

Broodstock Animals that are selected to be parents for the next generation

Candidate gene A gene suspected a priori to be underlying a particular trait

Cell The basic units of all living organisms, capable of independent proliferation and 10–100 micrometers in size

Centimorgan A measurement of genetic distance where one centimorgan is equal to a 1% chance that a marker at one locus on a chromosome will be separated form a marker at a second locus due to crossing over in a single generation

Centromere The constricted region on the chromosome where it attaches to the spindle during cell division

Challenge test Testing and recording of the response of organisms to a defined environmental challenge, e.g. survival rate or survival time after exposure to a specific pathogen

Chromatid A daughter chromosome still joined to its pair at the centromere

Chromosome A DNA thread, carrying genetic information and condensed into a physical structure located in the cell nucleus and visible under a microscope at cell cleavage

Combined selection The simultaneous use of records from an individual itself, its relatives and their pedigree relationships for selection

Connectedness Genetic connections (relationships) that tie together records from individuals in different test environments or year classes (e.g. sibs or progeny)

Control A standard used for comparison or an experiment established as a check of other experiments

Correlated response The degree to which selection for one trait will influence the value of another trait

Correlation A value expressing the degree of mutual relationship between two traits (zero for no relationship, one for complete dependence)

Cortisol A hormone produced in the adrenal gland

Covariation The degree of association between two traits

Crossbreeding The mating of animals derived from different species, strains (breeds) or lines

Cumulative mortality The number or proportion of animals that have died during a challenge test added over time

Cytosine A base found in DNA and RNA

Dam A female parent

Diploid cells Cells which have two sets of chromosomes (2 N), one inherited from each parent

Dissemination of genetic gain The transfer of selectively bred (genetically improved) animals from the breeding station to the industry and the use of those animals, or their progeny, by the industry for production purposes

DNA (deoxyribonucleic acid) The genetic material in all living organisms, except RNA viruses, consisting of four bases, adenine (A), cytosine (C), guanine (G) and thymine (T)

DNA pooling A QTL mapping technique to reduce genotyping requirements where DNA samples are pooled from extreme individuals of the phenotypic distribution

Domestication The genetic selection process that adapts wild animals to environmental conditions in captivity

Dominance Form of genetic expression where the effect of one allele (the dominant allele) masks the effect of other, recessive allele(s) at the same locus

Effective population size (Ne) The number of breeding animals in an idealised population (1:1 sex ratio among parents, equal contributions of progeny) that will give the same increase of inbreeding per generation as in the population in question

Embryo An organism in the early stages of development

Energy retention The energy stored in the body as a proportion of the total energy consumed

Environmental correlation The correlation between two different traits measured in the same animal caused by common environmental effects. Also describes the correlation when the same trait is measured in different individuals, caused by individuals sharing the same environment

Enzymes Proteins that control the rate of biochemical reactions in organisms

Epistasis The interaction between genes such that a gene at one locus alters the effect of genes at other loci

Factorial mating design Controlled and simultaneous fertilisation of eggs from each female with semen from several males and semen from each male with eggs from several females

Family selection Selection based on the average performance of full- or half-sib groups

FCE (feed conversion efficiency, feed efficiency) kg growth per kg feed consumed

FCR (feed conversion ratio) kg feed consumed per kg growth

Fecundity The number of eggs a species releases, also known as fertility

Feed conversion efficiency see FCE

Feed conversion ratio see FCR

Feed efficiency see FCE

Fertility Ability to produce gametes (sperm or eggs), fecundity

Fertilisation Union of two gametes to produce a zygote

First polar body The smaller of the two daughter cells resulting from first meiosis division of the primary oocyte

Fitness The relative ability of an individual to survive and transmit its genes to the next generation

Full-sibs Individuals that have both a common father and a common mother

Gamete A haploid sex cell, sperm or egg

Gametogenesis Formation of gametes, sperm (spermatozoa) and egg (ovum), each containing only half the genetic material (one of each chromosome pair) of the individual

GAS Gene-Assisted Selection, a form of MAS where the functional mutation underlying a QTL is used in selection

Gene The unit of heredity, or a segment of the DNA molecule containing information that can be transcribed and translated into proteins

Gene frequency see allele frequency

Generation interval The average age in years of parents when their progeny are born

Genetic correlation The correlation caused by common genetic effects when measuring different traits in the same individual (e.g. caused by pleiotropy). Also describes the correlation for a trait measured in different related individuals caused by common inheritance

Genetic drift The chance fluctuation of allele frequencies from generation to generation in a closed population

Genetic gain The change in the value of a trait due to selection (caused by the change in allele frequencies in response to selection)

Genetic marker A heritable and observable variant of a gene or DNA sequence that may be associated with a particular gene or trait

Genetic variance The component of the phenotypic variance in a trait that is caused by genetic differences between animals

Genetics The science of heredity

Genotype The total effect of all genes affecting a given trait (biometric definition) or the genetic makeup of an individual indicating which alleles that occur at a given locus (Mendelian definition)

Genotype-environment interaction (G×E) When the relative genetic performance of individuals or genetic groups (e.g. sib families or populations) for a given trait changes from one test environment to another

Gonad The sex gland, ovary or testes

Guanine A base found in DNA and RNA

Gynogenesis A special form of reproduction where progeny only inherit chromosomes (genes) from the female parent

Half-sibs Animals that have one parent in common

Haploid cells Cells (e.g. gametes) that have one set of chromosomes (1N) (see diploid)

Herbivorous Animals that feed only on plants

Heredity The genetic characteristics inherited by an individual from its parents

Heritability The proportion of the phenotypic variation between individuals for a given trait in a defined population and test environment that is due to genetic variation

Hermaphrodite Animals that have both male and female reproductive organs

Heterosis The increase in performance of crosses between strains or genetic lines above the average of the parent stocks

Homozygote When the same allele occurs in both gene copies of a locus in an animal's genotype

Inbreeding The mating of animals sharing common ancestry (related animals)

Inbreeding coefficient The probability that the two alleles occurring at one locus of an individual both descend from the same ancestor

Inbreeding depression The reduction in performance caused by inbreeding

Indirect selection Selection for one trait in order to obtain response in another trait

Individual selection Selection of animals based only on their own performance

LDLA Linkage Disequilibrium Linkage Analysis, a QTL mapping strategy that uses both linkage and LD information to determine the probability of a QTL being present at a particular site in the geneome

LD-MAS Linkage Disequilibrium Marker-Assisted Selection, a form of MAS using markers in population-wide LD with a QTL

LE-MAS Linkage Equilibrium Marker-Assisted Selection, a form of MAS using within-family linkage of markers to QTL

Life cycle Series of developmental phases of an organism from the zygote stage to reproduction and death

Linkage disequilibrium (LD) The non-random association of alleles between two loci

Linkage map A genetic map of a species that shows the position of genes or genetic markers relative to each other in terms of recombination frequency

Lipid Fat together with other organic compounds, not soluble in water

Locus A specific place on the chromosome where a gene is located, plural – loci

Maintenance requirement The amount of nutrients required to maintain the body tissues of an animal in the absence of physical activity

MAS Marker-Assisted Selection, the use of genetic markers linked to QTL affecting important traits in selection

Mass selection Individual selection

Mass spawning The simultaneous mixed spawning of a group of males and/or females in a common environment (e.g. the same tank)

Mating design The plan for systematic mating of selected animals

Meiosis The process of cell division where chromosomes in diploid germ cells are replicated and separated into haploid gametes

Metabolism All of the physical and chemical processes which take place in a living organism

Microsatellite A highly polymorphic DNA marker that consists of repeated sequences of usually two to five nucleotides

Migration The introduction of individuals from one breeding population into another

Mitosis Cell division where diploid chromosomes are replicated and passed to diploid daughter cells

Molecular genetics The study of and use of the DNA code for selective breeding or other purposes (e.g. marker-assisted selection)

Molluscs A group of animals, normally producing a protective shell, that includes mussels, oysters, scallops, abalone and clams

Monoculture The separate culture of different species

Morgan A unit of recombinant frequency for measuring genetic linkage

Multiplier A farm cooperating with the breeding station to multiply and disseminate the improved stock to the industry

Mutation A sudden heritable change in the DNA code

Natural selection A genetic process that adapts populations to their evironment through the increased reproductive success of individuals with improved fitness

Ne see effective population size

Nested mating design Controlled and simultaneous mating where semen from each male is used to fertilise eggs from a sample of two or more separate females (or vice versa) to produce full-sib families nested within half-sib families

Non-additive effect When the combined effects of alleles within or across loci deviate from the sum of their additive effects (e.g. dominance)

Normal distribution The most commonly found distribution of biological data where most observations cluster around the mean, dropping off to two long tails on either side

Nucleus (cell) A part of the cell, separated by a membrane, containing the chromosomes

Nucleus (breeding unit) The population of individuals that are candidates to be selected as parents to the next generation in a breeding program

Omnivorous Animals that feed both on animal and plant material

Oocyte A diploid cell that becomes an ovum after meiosis

Ovulation Release of ripe eggs from the ovary

Ovum An unfertilised egg cell

Ovary The female reproductive gland

Pedigree A diagram or matrix showing the genetic relationship between family members

Pedigree selection The selection of animals based on the performance of their parents, grandparents and/or other relatives

Phenotype The observed appearance or performance of an individual

Phenotypic correlation The correlation between phenotypic observations of different traits in the same individual

Phenotypic variance The total variance estimated for a phenotypic trait

PIT tag Passive Integrated Transponder tag, an electronic tag for identification purposes, usually inserted into the body cavity

Pleiotropy When a particular gene has an effect on several traits

Polar body The smaller of the two daughter cells resulting from meiosis division of the oocyte

Polyculture The mixed culture of different species in the same farm unit

Polyploid When organisms have more than two sets of chromosomes

Population A separate group of animals within a species

Population genetics The study of the allelic composition of populations

Progeny testing The testing of the performance of progeny for the selection among their parents

Protein Large molecules composed of sequences of amino acids

Protein retention The protein that is retained in the body as a proportion of the total amount of protein consumed by the individual

Purebreeding Mating of animals within a defined population

Qualitative trait A trait controlled by only one or a few genes

Quantitative genetics The theory of variation in traits caused by the simultaneous action of a large number of genes

Quantitative trait A trait controlled by many genes and showing continuous variation

Quantitative trait locus (QTL) A region of DNA harbouring a gene or genes with a detectable effect on a particular phenotypic trait

Random sample When every individual in the population has an equal and independent chance of being chosen for a sample

Recapture frequency Proportion of tagged/marked animals released to open sea or fresh waters and captured at a later stage in life, usually at spot of release

Recessive allele An allele whose effect is masked in the presence of a dominant allele and must be present in homozygote form to have an effect on a phenotype

Recombination The formation of new combinations of gene alleles by crossing over or independent assortment

Regression A measure of the dependency between two variables

Relationship The proportion of the genetic material shared by two individuals as a result of inheritance, or the classification of individuals according to their common ancestry (e.g. sibs, progeny)

Repeated mating When breeding animals produce progeny in several cycles, sometimes in consecutive generations

Response to selection The extent to which selection changes a trait in a population

RNA (ribonucleic acid) A single-stranded form of nucleic acid consisting of four bases, adenine (A), cytosine (C), guanine (G) and uracil (U)

Sample A restricted number of individuals chosen from a larger group according to a rule (often at random)

Sea ranching Production of fry/fingerlings that are released to the ocean, allowed to grow and then recaptured (in some cases, as they return to spot of release)

Second polar body The smaller of the two daughter cells resulting from the second meiosis division of the secondary oocyte

Secondary oocyte An egg cell after the shedding of the first polar body

Segregation The separation of parental chromosomes at meiosis

Selection The choice of parent broodstock from a larger group of candidates based on their breeding value, to produce a new generation with improved performance

Selection differential The difference between the population average and the average of the selected animals for a particular trait or breeding goal

Selection index A method for computing breeding values of individuals by weighing together information about different traits and from different sources (relatives)

Selection intensity Expresses the portion of the breeding candidates that are used as parents for the next generation

Selection method The method used for selecting breeders

Selection response The effect selection has on progeny

Selective breeding A plan for selecting and mating breeders

Self-fertilisation Fertilisation of an egg with sperm from the same individual

Selfing see self-fertilisation

Semen A fluid released by males containing sperm

Sex chromosomes A pair of chromosomes that affects the sex of an individual, for example female XX and male XY

Sexual maturation When females and males are ready to release mature eggs or sperm

Shellfish Molluscs and shrimp

Sire A male parent

SNP Single Nucleotide Polymorphism, genetic variation in a DNA sequence that occurs when a single nucleotide in a genome is altered

Spawning When aquatic species release mature eggs or sperm from the gonads

Species A group of animals that are freely able to breed with one another but are unable to freely breed with other species in the wild

Sperm A male gamete

Spermatocyte A diploid cell that undergoes meiosis to form sperm

Standard deviation A statistical term measuring the degree of variation of a trait, symbolized by σ (variance)

Strain A population of individuals with a common genetic history and often some common characteristics, sometimes called a breed in farmed species

Tandem selection The repeated selection for one trait in one generation and for another trait in the next generation

Test station A farm that receives tagged and pedigreed animals from a breeding station for testing under common farming conditions and supplies records of their performance to the breeding program

Tetraploids Individuals with four sets of chromosomes

Thymine A base found in DNA

Trait A character or state that may be recorded in an individual (e.g. growth rate, flesh tenderness, FCR, disease resistance)

Triploids Individuals with three sets of chromosomes

Truncation selection When all individuals with breeding values above a certain limit are selected as parents to the next generation

Uracil A base found in RNA

Variance A measure of variation, symbolised by σ^2 and calculated as the mean of the squares of individual deviations from the mean

Variation Differences between individuals in certain traits or characters, caused by genetic and environmental effects

Virus A non-cellular parasite, smaller than bacteria

Within-family selection Selection of the best performing individuals within each full-sib family, often used when families are tested in separate units with unknown differences in environmental conditions

YY male A male that is homozygote for the male sex chromosome, unlike the normal XY males in most species

Zygote A diploid cell formed by the union of two gametes.

References

Aguirre-Hernandez, J. and Sargan, D.R. 2005. Evaluation of candidate genes in the absence of positional information: A poor bet on a blind dog! Journal of Heredity, 96: 475–484.

Andersen, B.B. 1977. Genetic studies concerning growth rate in cattle, body development and feed efficiency (Genetiske undersøkelser vedrørende kvægets tilvækst, kropsudvikling og foderudnyttelse). Rep-No. 488, Natl. Inst. Anim. Sci., Copenhagen, 137pp.

Argue, B.J., Arce, S.M., Lotz, J.M. and Moss, S.M. 2002. Selective breeding of pacific white shrimp (Lito*penaeus vannamei*) for growth and resistance to Taura Syndrome. Aquaculture, 204: 447–460.

Austreng, E. and Refstie, T. 1979. Effect of varying dietary protein level in different families of rainbow trout. Aquaculture, 18: 145–156.

Ayles, G. and Baker, R. 1983. Genetic differences in growth and survival between strains and hybrids of rainbow trout (*Salmo gairdneri*) stocked in aquaculture lakes in the Canadian prairies. Aquaculture, 33: 269–280.

Bakos, J. 1979. Crossbreeding Hungarian races of common carp to develop more productive hybrids. In: Pillay, T.V.R. and Dill, W.A. (eds.). Advances in Aquaculture. Fishing Newes Books Ltd., Farnham, Surrey, pp. 633–635.

Baranski, M., Loughnan, S., Austin, C.M. and Robinson, N. 2006. A microsatellite linkage map of the blacklip abalone, *Haliotis rubra*. Animal Genetics, 37: 563–570.

Baranski, M., Rourke, M., Loughnan, S., Hayes, B., Austin, C. and Robinson, N. 2008. Detection of QTL for growth rate in the blacklip abalone (*Haliotis rubra* Leach) using selective DNA pooling. Animal Genetics, 39: 606–614.

Barber, B., Davis, C. and Hawes, R. 1998. Genetic improvement of oysters for the marine aquaculture industry. Abstract of the first Annual Northwest Aquaculture Conference and Exposition, Nov. 1998: 35pp.

Barlow, R. 1983. Benefit-cost analyses of genetic improvement program for sheep, beef cattle and pigs in Ireland. PhD. Thesis, University of Dublin. Ref. by Cunningham, E.P. 1983. Present and future perspective in animal breeding research. XV. International Congress of Genetics, New Delhi, India, 12–21 December, 19pp.

Beilharz, R.G., Luxford, B.G. and Wilkinson, J.L. 1993. Quantitative genetics and evolution: Is our understanding of genetics sufficient to explain evolution? Journal of Animal Breeding and Genetics, 100: 161–170.

Benjamin, A.T. and Sandra, J.P. 1996. Detection of obesity QTLs on mouse chromosomes 1 and 7 by selective DNA pooling. Genomics, 34: 389–398.

Bentsen, H.B. 1994. Genetic effects of selection on polygenetic traits with examples from Atlantic salmon, *Salmo salar L.* Aquaculture and Fisheries Management, 25: 89–102.

Bentsen, H.B. 2005. Polygenic inheritance. In: Gjedrem, T. (ed.). Selection and Breeding Programs in Aquaculture. Springer, Berlin, Heidelberg, 364pp.

Bentsen, H.B. and Olesen, I. 2002. Designing aquaculture mass selection programs to avoid high inbreeding rates. Aquaculture, 204: 349–359.

Bentsen, H.B., Eknath, A.E., Pallada-de Vera, M.S., Danting, J.C., Bolivar, H.L., Reyes, R.A., Dionisio, E.E., Longalong, F.M., Circa, A.V., Tayamen, M.M. and Gjerde, B. 1998. Genetic improvement of farmed tilapias: Growth performance in a complete diallel cross experiment with eight strains of *Oreochromis niloticus*. Aquaculture, 160: 145–173.

Bentsen, H.B., Eknath, A.E., Rye, M., Thodesen, J. and Gjerde, B. 2003. Genetic improvement of farmed tilapias. Response to selection for growth performance in the GIFT project. International Association for Genetics in Aquaculture VIII, 9–15 November, Puerto Varas, Chile: 68.

Bilio, M. 2007–2008. Controlled reproduction and domestication in aquaculture. Aquaculture Europe. Part I, 32(1): 14 pp; Part II, 32(3): 23 pp; Part III, 33(1): 19 pp and Part IV, 33(2): 24pp.

Bjerkeng, B. 2008. Carotenoids in aquaculture: Fish and crustaceans. Carotenoids, 4: 237–254.

Bolivar, R. 1999. Estimation of response to within-family selection for growth rate in tilapia. Diss. Abst. Int. Pt. B-Sci. and Eng., 60(3): 934.

Bolivar, R.B. and Newkirk, G.F. 2002. Response to within-family selection for body weight in Nile tilapia (*Oreochromis niloticus*) using a single trait animal model. Aquaculture, 204: 371–381.

Bondary, K. 1983. Response to bidirectional selection for body weight in channel catfish. Aquaculture, 33: 73–81.

Bovenhuis, H. and Spelman, R.J. 2000. Selective genotyping to detect quantitative trait loci for multiple traits in outbred populations. Journal of Dairy Science, 83: 173–180.

Brown, R.C., Woolliams, J.A. and McAndrew, B.J. 2005. Factors influencing effective population size in commercial populations of gilthead sea bream, *Sparus aurata*. Aquaculture, 247: 219–225.

Bulmer, M.G. 1971. The effect of selection on genetic variability. The American Naturalist, 105: 201–221.

Calhoon, R.E. and Bohren, B.B. 1974. Genetic gains from reciprocal recurrent and within-line selection for egg production in the fowl. Theoretical and Applied Genetics, 44: 364–372.

Carlin, B. 1969. Salmon tagging experiments. Laksforskningsinstitutet (Swedish Salmon Research Institute). Meddelanden, 2–4: 8–13.

Chapman, A.B. 1962. History of the genetics of quantitative variability. Lecture Notes, 236pp.

Charo-Karisa, H., Rezk, M.A., Bovenhuis, H. and Komen, H. 2005. Heritability of cold tolerance in Nile tilapia, *Oreochromis niloticus*, juveniles. Aquaculture, 249: 115–123.

Cherfas, N.B. 1981. Gynogenesis in fishes. In: Kirpichnikov, V.S. (ed.). Genetic Bases of Fish Selection. Springer-Verlag, Berlin, Heidelberg, New York, 412pp.

Chevassus, B. 1979. Hybridization in salmonids: Results and perspectives. Aquaculture, 17: 113–128.

Chistiakov, D.A., Hellemans, B. and Volckaert, F.A.M. 2005b. Microsatellites and their genomic distribution, evolution, function and applications: A review with special reference to fish genetics. Aquaculture, 255: 1–29.

Chistiakov, D.A., Hellemans, B., Haley, C.S., Law, A.S., Tsigenopoulos, C.S., Kotoulas, G., Bertotto, D., Libertini, A. and Volckaert, F.A.M. 2005a. A microsatellite linkage map of the European sea bass *Dicentrarchus labrax* L. Genetics, 170: 1821–1826.

Chourrot, D. 1984. Pressure-induced retention of second polar body and suppression of first cleavage in rainbow trout: Production of all-triploids, all-tetraploids, and heterozygous diploid gynogentics. Aquaculture, 36: 111–126.

Chourrout, D. 1980. Thermal induction of diploid gynogenesis and triploidy in the eggs of the rainbow trout (*Salmo gairdneri Richardson*). Reproduction Nutrition Development, 20(3A): 727–733.

Cock, J., Gitterle, T., Salazar, M. and Rye, M. 2008. Breeding for disease resistance of Penaeid shrimps. Aquaculture, 286:1–11.

Coimbra, M.R.M., Kobayashi, K., Koretsugu, S., Hasegawa, O., Ohara, E., Ozaki, A., Sakamoto, T., Naruse, K. and Okamoto, N. 2003. A genetic linkage map of the Japanese flounder, *Paralichthys olivaceus*. Aquaculture, 220: 203–218.

Comstock, R.E., Robinson, H.F. and Harvey, P.H. 1949. A breeding procedure designed to make use of both general and specific combining ability. Agronomy Journal, 41: 360–367.

Crenshaw, J.W.Jr., Heffernan, P.B. and Walker, R.L. 1991. Heritability for growth rate in the southern bay scallop. *Argopecten irradians concentricus* (Say. 1822). Journal of Shellfish Research, 10(1): 55–63.

Danzmann, R.G., Jackson, T.R. and M. Ferguson, M. 1999. Epistasis in allelic expression at upper temperature tolerance QTL in rainbow trout. Aquaculture, 173: 45–58.

Darvasi, A. and Soller, M. 1992. Selective genotyping for determination of linkage between a marker locus and a quantitative trait locus. Theoretical Appleid Genetics, 85: 353–359.

Darvasi, A. and Soller, M. 1994. Selective DNA pooling for determination of linkage between a molecular marker and a quantitative trait locus. Genetics, 138: 1365–1373.

Davis, G.P. and Hetzel, D.J.S. 2000. Integrating molecular genetic technology with traditional approaches for genetic improvement in aquaculture species. Aquaculture Research, 31: 3–10.

Dekkers, J.C.M. 2004. Commercial application of marker- and gene-assisted selection in livestock: Strategies and lessons. Journal of Animal Science, 82: E313–328.

Devlin, R.H., Biagi, C.A., Yesaki, T.Y., Smailus, D.E. and Byatt, J.C. 2001. Griwth of domesticated transgenic fish. Nature, 409: 781-782.

Dickerson, G.E. 1952. Inbred lines for heterosis tests? In: Gowen, J.W. (ed.). Heterosis. Iowa State College, Ames, Iowa, pp. 330–351.

Dickerson, G.E. 1978. Animal size and efficiency – basic concepts. Animal Production, 27: 367–379.

Donaldson, R.L. 1968. Selective breeding of salmonid fishes. In: McNeil, W. (ed.). Marine Aquaculture, 23–24 May. Oregon State University Press, Corvallis, OR, pp. 65–74.

Doyle, R.W. and Talbot, A.J. 1986. Artificial selection on growth and correlated selection on competitive behaviour in fish. Canadian Journal of Fisheries and Aquatic Sciences, 43: 1059–1064.

Dunham, R.A. 1987. American catfish breeding programs. In: Tiewes, K. (ed.). Selection, Hybridization and Genetic Engineering in Aquaculture. Schriften der Bundesforschungsanstalt für Fisherei, Vol II. Hamburg, pp. 407–416.

Dunham, R.A. 2006. Comparison of six generations of selection, interspecific hybridization, intraspecific crossbreeding and gene transfer for growth improvement in Ictalurus catfish. IAGA, 26–30 June, Montpellier, Abstract: 22.

Dunham, RA. 2004. Aquaculture and fisheries biotechnology: genetic approaches. CABI Publishing, Cambridge, MA, 372pp.

Dupond-Nivet, M., Vandeputte, M., Vergnet, A., Merdy, O., Haffray, P., Chavanne, H. and Chatain, B. 2008. Heritabilities and GxE interactions for growth in the European sea bass (*Dicentrarchus labrax L.*) using a marker-based pedigree. Aquaculture, 275: 81–87.

Edwards, D.J. and Gjedrem, T. 1979. Genetic variation in survival of brown trout eggs, fry and fingerlings in acidic water. SNSF – project, Norway, FR 16/79, 28pp.

Edwards, D.J., Austreng, E., Risa, S. and Gjedrem, T. 1977. Carbohydrate in rainbow trout diets. I. Growth of fish of different families fed diets containing different proportions of carbohydrate. Aquaculture, 11: 31–38.

Eknath, A.E., Bentsen, H.B., Ponzoni, R.W., Rye, M., Nguyen, N.H., Thodesen, J. and Gjerde, B. 2007. Genetic improvement of farmed tilapias: Composition and genetic parameters of a synthetic base population of *Oreochromis niloticus* for selective breeding. Aquaculture, 273: 1–14.

Eknath, A.E., Dey, M.M., Rye, M., Gjerde, B., Abella, T.A., Sevilleja, R., Tayamen, M.M., Reyes, R.A. and Bentsen, H.B. 1998. Selective breeding of Nile tilapia for Asia. Proceedings of the 6th World Congress on Genetics Applied to Livestock Production, 27: 89–96.

Eknath, A.E., Tayamen, M.M., Palada-de Vera, M.S., Danting, J.C., Reyes, R.A., Dinoisio, E.E., Capilli, J.B., Bolivar, H.L., Abella, T.A., Circa, A.V., Bentsen, H.B., Gjerde, B., Gjedrem, T. and Pullin, R.S.V. 1993. Genetic improvement of farmed tilapias: The growth performance of eight strains of *Oreochromis niloticus* tested in different farm environments. Aquaculture, 111: 171–188.

Embody, G.C. and Hyford, C.D. 1925. The advantage of rearing brook trout fingerlings from selected breeders. Transaction of the American Fisheries Society, 55: 135–138.

Enfield, F.D. 1979. Long term effects of selection; the limits to response. In proceedings of a symposium on "Selection experiments in laboratory and domestic animals". Commonwelth Agric. Beureaux, pp. 69–86.

Eudeline, B., Allen, S.K.Jr. and Guo, X. 2000. Optimization of tetraploid induction in Pacific oysters, *Crassotrea gigas*, using first polar body as a natural indicator. Aquaculture, 187: 73–84.

Evans, B., Bartlett, J., Sweijd, N., Cook, P. and Elliott, N.G. 2004. Loss of genetic variation at microsatellite loci in hatchery produced abalone in Australia (*Haliotis rubra*) and South Africa (*Haliotis midae*). Aquaculture, 233: 109–127.

Evans, S. and Langdon, C. 2006. Direct and indirect responses to selection on individual body weight in the pacific oyster (*Crassostrea gigas*). Aquaculture, 261: 546–555.

Falconer, D.S. 1960. Introduction to Quantitative Genetics. Oliver and Boyd, Edinburgh, 365pp.

Falconer, D.S. and Mackay, T.F.C. 1996. Introduction to Quantitative Genetics, Longman, ISBN 0582-24302-5, 464pp.

FAO yearbook. 2007. Fisheries statistics. Aquaculture Production. Vol. 100/2, 2005. 206pp.

Farnir, F., Grisart, B., Coppieters, W., Riquet, J., Berzi, P., Cambisano, N., Karim, L., Mni, M., Moisio, S., Simon, P., Wagenaar, D., Vilkki, J. and Georges, M. 2002. Simultaneous mining of linkage and linkage disequilibrium to fine map quantitative trait loci in outbred half-sib pedigrees: Revisiting the location of a quantitative trait locus with major effect on milk production on bovine chromosome 14. Genetics, 161: 275–287.

Fevolden, S.E., Nordmo, R., Refstie, T. and Røed, K.H. 1993. Disease resistance in Atlantic salmon (*Salmo salar*) selected for high or low responses to stress. Aquaculture, 109: 215–224.

Fevolden, S.E., Refstie, T. and Røed, K.H. 1992. Disease resistance in rainbow trout (*Oncorhynchus mykiss*) selected for stress response. Aquaculture, 104: 19–29.

Fevolden, S.E., Røed, K.H. and Fjalestad, K.T. 2002. Selection response of cortisol and lysozyme in rainbow trout and correlation to growth. Aquaculture, 205: 61–75.

Fimland, E. 1979. The effect of selection on additive genetic parameters. Z. Tierz., Zuchtungsbiol., 96: 120–134.

Fjalestad, K.T. 2005. Breeding strategies. In: Gjedrem, T. (ed.). Selection and Breeding Programs in Aquaculture. Springer, Berlin, Heidelberg, 364pp.

Fjalestad, K.T., Gjedrem, T., Carr, W.H. and Sweeney, J.N. 1997. Final report: The Shrimp Breeding Program Selective Breeding of *Penaeus vannamei*. AKVAFORSK, Rep. no. 17/97, 85pp.

Fletcher, G.L., Shears, M.A., Yaskowiak, E.S., King, M.J. and Goddard, S.V. 2004. Gene transfer: potential to enhance the genome of Atlantic salmon for aquaculture. Australian Journal of Experimental Agriculture, 44: 1095–1100.

Flynn, F.M.O., Baily, J.K. and Friars, G.W. 1999. Response to two generations of index selection in Atlantic salmon (*Salmo salar*). Aquaculture, 173: 143–147.

Folkestad, A., Wold, J.P., Rørvik, K.A., Tschudi, J., Haugholt. K.H., Kolstad, K. and Mørkøre, T. 2008. Rapid and non-invasive measurements of fat and pigment concentrations in live and slaughtered Atlantic salmon (*Salmo salar L.*). Aquaculture, 280: 129–135.

Friars, G.W., Bailey, J.K. and Sounders, R.L. 1979. Considerations of a method of analysing diallel cross of Atlantic salmon. Canadian Journal of Genetics and Cytology, 21: 121–128.

Friars, G.W., McMillan, I.M., Quinton, V.M., O'Flynn, F.M., McGeachy, S.A. and Benfey, T.J. 2001. Family differences in relative growth of diploid and triploid Atlantic salmon (*Salmo salar L.*). Aquaculture, 192: 23–29.

Gall, G.A.E. 1975. Genetics of reproduction in domesticated rainbow trout. Journal of Animal Science, 40: 19–28.

Gall, G.A.E. and Huang, N. 1988a. Heritability and selection schemes for rainbow trout: Body weight. Aquaculture, 73: 43–56.

Gall, G.A.E. and Huang, N. 1988b. Heritability and selection for rainbow trout: Female reproductive performance. Aquaculture, 73: 57–66.

Gall, G.A.E., Bakar, Y. and Famula, T. 1993. Estimating genetic change from selection. Aquaculture, 111: 75–88.

Gardner, E.J. and Snustad, D.P. 1981. Principles of Genetics. John Wiley and Sons, New York, Chichester, Brisbane, Toronto, Singapore, 611pp.

Gardner, E.J., Simmons, M.J. and Snustad, D.P. 1991. Principles of Genetics (Eighth Edition) John Wiley & Sons Inc., 740pp.

Georges, M., Nielson, D., Mackinnon, M., Mishra, A., Okimoto, R., Pasquino, A.T., Sargent, L.S., Sorensen, A., Steele, M.R., Zhao, X., Womack, J.E. and Hoeschele, I. 1995. Mapping quantitative trait loci controlling milk production in dairy cattle by exploiting progeny testing. Genetics, 139: 907–920.

Gharbi, K., Gautier, A., Danzmann, R.G., Gharbi, S., Sakamoto, T., Hoyheim, B., Taggart, J.B., Cairney, M., Powell, R., Krieg, F., Okamoto, N., Ferguson, M.M., Holm, L.-E. and Guyomard, R. 2006. A Linkage Map for brown trout (*Salmo trutta*): Chromosome homeologies and comparative genome organization with other salmonid fish. Genetics, 172: 2405–2419.

Gilbey, J., Verspoor, E., McLay, A. and Houlihan, D. 2004. A microsatellite linkage map for Atlantic salmon (*Salmo salar*). Animal Genetics, 35: 98–105.

Gitterle, T., Johansen, H., Erazo, C., Lozano, C., Cock, J., Salazar, M. and Rye, M. 2006. Response to multi trait selection for harvest weight, overall survival, and resistance to white spot syndrome virus (WSSV) in *Penaeus (Litopenaeus) vannamei.* IAGA, 26–30 June 2006, Montpelier, Abstract: 35.

Gitterle, T., Rye, M., Salte, R., Cock, J., Johansen, H., Lozano, C., Suárez, J.A. and Gjerde, B. 2005. Genetic (co)variation in harvest body weight and survival in *Penaeus vannamei* under standard farming conditions. Aquaculture, 243: 83–92.

Gjedrem, T. (ed.). 2005. Selection and Breeding Programs in Aquaculture. Springer, Berlin, Heidelberg, 364pp.

Gjedrem, T. 1967. Selection indexes compared with single trait selection. I. The efficiency of including correlated traits. Acta Agricultura Scandinavica, 17: 263–275.

Gjedrem, T. 1972. A study on the definition of the aggregate genotype in a selection index. Acta Agricultura Scandinavica, 22: 11–16.

Gjedrem, T. 1976. Genetic variation in tolerance of brown trout to acid water. SNSF-project, Norway, FR 5/76, 11pp.

Gjedrem, T. 1979. Farming of salmon and rainbow trout. (Oppdrett av laks og aure). Landbrukdforlaget, Oslo, 332pp.

Gjedrem, T. 1995. Genetics and breeding for aquaculture. (Genetikk og avlslære for akvakultur). Landbruksforlaget, 112 pp.

Gjedrem, T. 1997. Selective breeding to improve aquaculture production. World Aquaculture, 28(1): 33–45.

Gjedrem, T. 1998. Diseases and parasites in fish. Genetic improvement of resistance. GIFT, Final Report (March 1988 to Desember 1997). ICLARM publication: Attachment 11: 27pp.

Gjedrem, T. 2000. Genetic improvement of cold-water fish species. Aquaculture Research, 31: 25–33.

Gjedrem, T. 2004. Status for Breeding Programs in Aquaculture. Fish Breeder's Roundtable 2004, Håholmen, Norway.

Gjedrem, T. and Andersen, Ø. 2005. Basic genetics. In: Gjedrem, T. (ed.). Selection and Breeding Programs in Aquaculture. Springer, Berlin, Heidelberg, 364pp.

Gjedrem, T. and Fjalestad, K.T. 1997. Body Weight Distribution of Atlantic Salmon Parr During Domestication. AKVAFORSK, manuscript, 14pp.

Gjedrem, T. and Gjøen, H.M. 1995. Genetic variation in susceptibility of Atlantic salmon, *Salmo salar L.*, to furunculosis, BKD and cold water vibriosis. Aquaculture Research, 26: 129–134.

Gjedrem, T. and Olesen, I. 2005. Basic statistical parameters. In: Gjedrem, T. (ed.). Selection and Breeding Programs in Aquaculture. Springer, Berlin, Heidelberg, 364pp.

Gjedrem, T. and Thodesen, J. 2005. Selection. In: Gjedrem, T. (ed.). Selection and Breeding Programs in Aquaculture. Springer, Berlin, Heidelberg, 364pp.

Gjedrem, T., Gjøen, H.M. and Gjerde, B. 1991a. Genetic origin of Norwegian farmed Atlantic salmon. Aquaculture, 98: 41–50.

Gjedrem, T., Gjøen, H.M., Hardjamulia, A., Sudarto, Ir., Widiaty, A., Gustiano, R., Kristanto, A.H., Emmawati, L. and Hadie, W. 1997. Breeding plan for Nile tilapia in Indonesia: Individual (mass) selection. INGA, ICLARM, Report No. 4., 11pp.

Gjedrem, T., Salte, R. and Gjøen, H.M. 1991b. Genetic variation in susceptibility of Atlantic salmon to furunculosis. Aquaculture, 97: 1–6.

Gjerde, B. 1984. Response to individual selection for age at sexual maturity in Atlantic salmon. Aquaculture, 38: 229–240.

Gjerde, B. 1986. Growth and reproduction in fish and shellfish. Aquaculture, 57: 37–55.

Gjerde, B. 1987. Predicting carcass composition of rainbow trout by computerized tomography. Journal of Animal Breeding and Genetics, 104: 121–136.

Gjerde, B. 1988. Complete diallel cross between six inbred groups of rainbow trout, *Salmo gairdneri*. Aquaculture, 75: 71–87.

Gjerde, B. 1991. Breeding theory for salmon and rainbow trout. (Avlslære. Laks og regnbogeaure). Lecturenotes, Landbruksbokhandelen, Agricultural University of Norway, ISBN 82-557-0374-8. 155pp.

Gjerde, B. 2005a. Prediction of breeding values. In: Gjedrem, T. (ed.). Selection and Breeding Programs in Aquaculture. Springer, Berlin, Heidelberg, 364pp.

Gjerde, B. 2005b. Design of Breeding Programs. In: Gjedrem, T. (ed.). Selection and Breeding Programs in Aquaculture. Springer, Berlin, Heidelberg, 364pp.

Gjerde, B and Gjedrem, T. 1984. Estimates of phenotypic and genetic parameters for carcass traits in Atlantic salmon and rainbow trout. Aquaculture, 36: 97–110.

Gjerde, B. and Korsvoll, A. 1999. Realized selection differentials for growth rate and early sexual maturity in Atlantic salmon. Abstracts, Aquaculture Europe 99, Trondheim, Norway, August 7–10: 73–74.

Gjerde, B. and Olsen, B. 1990. Economic value of breeding program. (Økonomisk verdi av avlsarbeidet). Husdyrforsøksmøtet 1990, Akvakultur. Aktuelt fra Statens Fagtjeneste for Landbruket, Nr.5: 61–65.

Gjerde, B. and Refstie, T. 1984. Complete diallel cross between five strains of Atlantic salmon. Livestock Production Science, 11: 207–225.

Gjerde, B. and Schaeffer, L.R. 1989. Body traits in rainbow trout. II. Estimates of heritabilities and phenotypic and genetic correlations. Aquaculture, 80: 25–44.

Gjerde, B., Gjøen, H.M. and Villanueva, B. 1996. Optimum designs for fish breeding programs with constrained inbreeding. Mass selection for a normally distributed trait. Livestock Production Science, 47: 59–72.

Gjerde, B., Gunnes, K. and Gjedrem, T. 1983. Effect of inbreeding on survival and growth rate. Aquaculture, 34: 327–332.

Gjerde, B., Padala, V.G., Reddy, P.V.G.K., Mahapatra, K.D., Saha, J.N., Jana, R.K., Keher, P.K., Sahu, M., Lenka, S., Govindassamy, P. and Rye, M. 2002. Growth and survival in two complete diallele crosses with five stocks of Rohu (*Labeo rohita*). Aquaculture, 209: 103–115.

Gjerde, B., Pante, M.J.R. and Bæverfjord, G. 2005. Genetic variation for vertebral deformity in Atlantic salmon (*Salmo salar*). Aquaculture, 244: 77–87.

Gjerde, B., Rye, M. and Gjedrem, T. 2007a. State of the art in selective breeding of aquaculture species. Manuscript.

Gjerde, B., Simianer, H. and Refstie, T. 1994. Estimates of genetic and phenotypic parameters for body weight, growth rate and sexual maturity in Atlantic salmon. Livestock Production Science, 38: 133–143.

Gjerde, B., Sonesson, A.K., Storset, A. and Rye, M. 2007b. Selective breeding and genetics – Atlantic salmon. The Research Council of Norway. Aquaculture Reseach: From Cage to Consumer, pp. 268–284.

Gjerde, B., Terjesen, B.F., Barr, Y., Lein, I. and Thorland, I. 2004. Genetic variation for juvenile growth and survival in Atlantic cod (*Gadus morhua*). Aquaculture, 236: 167–177.

Gjedrem, T. 1985. Improvement of productivity through breeding scheme. Geo Journal, 10(3): 233–241.

Gjøen, H.M., Refstie, T., Ulla, O. and Gjerde, B. 1997. Genetic correlations between survival of Atlantic salmon in challenge and field tests. Aquaculture, 158: 277–288.

Gjøen, H.M., Storebakken, T., Austreng, E. and Refstie, T. 1993. Genotypes and nutrient utilization. Fish nutrition in Practise. Blarritz (France), June 24–27, 1991. INRA, Paris (Les Colleques, no. 61), 19–26pp.

Goddard, M. 1992. A mixed model for analyses of data on multiple genetic markers. Theoretical Applied Genetics, 83: 878–886.

Godin, D.M., Carr, W.H., Hagino, G., Segura, F., Sweeney, J.N. and Blankenship, L. 1996. Evaluation of a florecent elastomer internal tag in juvenile and adult shrimp *Penaeus vannamei*. Aquaculture, 139: 243–248.

Grammeltvedt, A.F. 1975. Chromosomes of salmon (*Salmo salar*) by leukocyte culture. Aquaculture, 5: 205–209.

Grisart, B., Coppieters, W., Farnir, F., Karim, L., Ford, C., Berzi, P., Cambisano, N., Mni, M., Reid, S., Simon, P., Spelman, R., Georges, M. and Snell, R. 2002. Positional candidate cloning of a QTL in dairy cattle: Identification of a missense mutation in the bovine DGAT1 gene with major effect on milk yield and composition. Genome Ressearch, 12: 222–231.

Gunnes, K. and Gjedrem, T. 1978. Selection experiments with salmon. IV. Growth of Atlantic salmon during two years in the sea. Aquaculture, 15: 19–33.

Gunnes, K. and Gjedrem, T. 1981. A genetic analysis of body weight and length in rainbow trout reared in sea water for 18 months. Aquaculture, 24: 161–174.

Guo, X. and Allen, S.K. 1994a. Sex determination and polyploidy gigantism in the Dwarf surfclam (*Mulinia lateralis Say*). Genetics, 138: 1199–1206.

Guo, X. and Allen, S.K. 1994b. Viable tetraploids in the Pacific oyster (*Crassostrea gigas* Thunberg) produced by inhibiting polar body I in eggs from triploids. Molecula Marine Biology and Biotechnology, 3: 42–50.

Guo, X., DeBrosse, G.A. and Allen, S.K. 1996. All-triploid Pacific oysters (*Crassostrea gigas* Thunberg) produced by mating tetraploids and diploids. Aquaculture, 142: 149–164.

Haard, N.F. 1992. Control of chemical composition and food quality attributes of cultured fish. Food Research International, 25: 289–307.

Haldane, J.B.S. 1919. The combination of linkage values, and the calculation of distance between the loci of linked factors. Journal of Genetics, 8: 299–309.

Hamm, D.E. and Burton, R.S. 2000. Population genetics of black abalone, *Haliotis cracherodii*, along the central California coast. Journal of Experimental Marine Biology and Ecology, 254: 235–247.

Hand, R.E., Nell, J.A. and Thompson, P.A. 2004. Studies on triploid oysters in Australia XIII. Performance of diploid and triploid Sydney rock oyster, *Saccostrea glomerata* (Gould, 1850), progeny from a third generation breeding line. Aquaculture, 233: 93–107.

Hara, M. and Sekino, M. 2005. Genetic difference between Ezo-awabi *Haliotis discus hannai* and Kuro-awabi *H. discus* populations: Microsatellite-based population analysis in Japanese Abalone. Fisheries Science, 71: 754–766.

Haugen Regnmark, A., Slettan, A., Skaala, Ø., Lie, Ø. and Lingaas, F. 2006. Genetic variability in wild and farmed Atlantic salmon (*Salmo salar*) strains estimated by SNP and microsatellites. Aquaculture, 253: 229–237.

Havenstein, G.B., Ferkel, P.R. and Qureshi, M.A. 2003. Growth, liveability, and feed conversion of 1957 versus 2001 broilers when fed 1957 and 2001 broiler diets. Poultry Science, 82: 1500–1508.

Havenstein, G.B., Ferkel, P.R., Scheideler, S.E. and Larson, B.T. 1994. Growth, liveability, and feed conversion of 1957 vs 1991 broilers when fed "typical" 1957 and 1991 broiler diets. Poultry Science, 73: 1785–1794.

Hayes, B., Sonesson, A.K. and Gjerde, B. 2005. Evaluation of three strategies using DNA markers for traceability in aquaculture species. Aquaculture, 250: 70–81.

Hayes, B.J., Gjuvsland, A. and Omholt, S. 2006. Power of QTL mapping experiments in commercial Atlantic salmon populations, exploiting linkage and linkage disequilibrium and effect of limited recombination in males. Heredity, 97: 19–26.

Hazel, L.N. 1943. The genetic basis for constructing selection indexes. Genetics, 28: 476–490.

Hazel, L.N. and Lush, J.L. 1942. The efficiency of three methods of selection. Journal of Heredity, 33: 393–399.

Henderson, C.R. 1975. Best linear unbiased estimation and prediction under a selection model. Biometrics, 31: 423–437.

Henryon, M., Jokumsen, A., Berg, P., Lund, I., Pedersen, P.B., Olesen, N.J. and Slierendrecht, W.J. 2002. Genetic variation for growth rate, feed conversion efficiency, and disease resistance exists within a farmed population of rainbow trout. Aquaculture, 209: 59–76.

Hershberger, W.K., Myers, J.M., Iwamoto, R.N., Mcauley, W.C. and Saxton, A.M. 1990. Genetic changes in the growth of coho salmon (*Onchorhynchus kisutch*) in marine net-pens, produced by ten years of selection. Aquaculture, 85: 187–197.

Hetzel, D.J.S., Crocos, P.J., Davis, G.P., Moore, S.S. and Preston, N.C. 2000. Response to selection and heritability for growth in the Kuruma prawn, *Penaeus japonicus*. Aquaculture, 181: 215–223.

Hill, W.G. 2008. Estimation, effectiveness and opportunities of long term genetic improvement in animals and maize. Lohmann Information, 43: 3–20.

Holt, M., Meuwissen, T. and Vangen, O. 2005. Long-term responses, changes in genetic variances and inbreeding depression from 122 generations of selection on increased litter size in mice. Journal of Animal Breeding and Genetics, 122: 199–209.

Holtsmark, M., Klemetsdal, G., Sonesson, A.K. and Woolliams, J.A. 2007. Synthetic base population: Exploring the effect of the intensity of selection in early stages of the selection program with a finite locus model. In: Design of base population and selection strategies for sustainable genetic improvement program for fish species. Dr. thesis at Norwegian University of Life Sciences, 69–96.

Holtsmark, M., Klemetsdal, G., Sonesson, A.K. and Woolliams, J.A. 2008. Establishing a base population for a breeding program in aquaculture, from multiple subpopulations, differentiated by genetic drift: II. Sensitivity to assumptions on the additive genetic relationships of base animals. Aquaculture, 274: 241–246.

Holtsmark, M., Sonesson, A.K., Gjerde, B. and Klemetsdal, G. 2006. Number of contributing subpopulations and mating design in the base population when establishing a selective breeding program for fish. Aquaculture, 258: 241–249.

Houston, R.D., Haley, C.S., Hamilton, A., Guy, D.R., Tinch, A.E., Taggart, J.B., McAndrew, B.J. and Bishop, S.C. 2008. Major QTL affect resistance to infectious pancreatic necrosis in Atlantic salmon (*Salmo salar*). Genetics, 178: 1109–1115.

Howe, A. and Kocher, T.D. 2003. Comparative mapping of QTL for red body color in tilapia. In: Final Abstracts Guide. Plant and Animal Genome Conference XI, January 11–15, 2003, San Diego, USA.

Huang, B.X., Peakall, R. and Hanna, P.J. 2000. Analysis of genetic structure of blacklip abalone (*Haliotis rubra*) populations using RAPD, minisatellite and microsatellite markers. Marine Biology, 136: 207–216.

Huang, S.S.O. and Liao, I.C. 1990. Response to mass selection for growth rate in *Oreochromis niloticus*. Aquaculture, 85: 199–205.

Hubert, S. and Hedgecock, D. 2004. Linkage maps of microsatellite DNA markers for the Pacific oyster *Crassostrea gigas*. Genetics, 168: 351–362.

Hulata, G., Wohlfarth, G.W. and Halevy, A. 1986. Mass selection for growth rate in the Nile tilapia (*Oreochromis niloticus*). Aquaculture, 57: 177–184.

Hussain, M.G., Islam, M.S., Hossain, M.A., Wahid, M.I., Kohinoor, A.H.M., Dey, M.M. and Mazid, M.A. 2002. Stock improvement of silver barb (*Barbodes gonionotus* Bleeker) through several generations of genetic selection. Aquaculture, 204: 469–480.

Hussain, M.G., Rao, G.P.S., Humayun, N.M., Randell, C.F., Penman, D.J., Keme, D., Bromage, N.R., Myers, J.M. and McAndrew, B.J. 1995. Comparative performance of growth, biochemical composition and endocrine profiles in diploid and triploid tilapia *Orechromis niloticus* L. Aquaculture, 138: 87–97.

Ibarra, A.M., Ramirez, J.L., Ruiz, C.A., Cruz, P. and Avila, S. 1999. Realized heritabilities and genetic correlation after dual selection for total weight and shell width in catarina scallop (*Argopecten ventricosus*). Aquaculture, 175: 227–241.

Ihara, N., Takasuga, A., Mizoshita, K., Takeda, H., Sugimoto, M., Mizoguchi, Y., Hirano, T., Itoh, T., Watanabe, T., Reed, K.M., Snelling, W.M., Kappes, S.M., Beattie, C.W., Bennett, G.L. and Sugimoto, Y. 2004. A comprehensive genetic map of the cattle genome based on 3802 microsatellites. Genome Research, 14: 1987–1998.

Jackson, T.R., Ferguson, M.M., Danzmann, R.G., Fishback, A.G., Ihssen, P.E., O'Connell, M. and Crease, T.J. 1998. Identification of two QTL influencing upper temperature tolerance in three rainbow trout (*Oncorhynchus mykiss*) half-sib families. Heredity, 80: 143–151.

Jarayabhand, P. and Tavornyuikarn, M. 1995. Realized heritability estimation on growth rate of oyster, *Succostrea cucullata* born 1778. Aquaculture, 138: 111–118.

Jensen, A., Bodvin, T., Bye, O., Hjelt, K.A., Holmefjord, I., Håstein, T., Olafsen, H., Pedersen, F., Reve, T., Rogne, J. and Standal, D. 1999. Norwegian possibilities for value creation within aquaculture. (Norske muligheter innen havbruk). Trondheim, 35pp.

Jentoft, S., Aastveit, A.H., Torjesen, P.A. and Andersen, Ø. 2006. Effects of stress on growth, cortisol and glucose levels in non-domesticated Eurasian perch (*Perca fluviatilis*) and domesticated rainbow trout (*Oncorhynchus mykiss*). Comparative Biochemistry and Physiology, 141: 353–358.

Johnson, M. and Griffiths, L. 2005. A genetic analysis of serotonergic biosynthetic and metabolic enzymes in migraine using a DNA pooling approach. Journal of Human Genetics, 50: 607–610.

Jonasson, J. 1984. Comparison of sexual maturation between diploid and triploid rainbow trout (*Salmo gairdneri*). Msc. Thesis, Agricultural University of Norway, 43pp.

Jonasson, J. 1993. Selection experiments in salmon ranching. I. Genetic and environmental sources of variation in survival and growth in freshwater. Aquaculture 109: 225–236.

Jonasson, J. 1994. Selection experiments in Atlantic salmon ranching. V. Realized response to selection for increased return rate of grilse. Phd thesis at Agricultural University of Norway, 15pp.

Jonasson, J., Gjerde, B. and Gjedrem, T. 1997. Genetic parameters for return rate and body weight of sea-ranched Atlantic salmon. Aquaculture, 154: 219–231.

Jonasson, J., Stefansson, S.E., Gudnasson, A. and Steinarsson, A. 1999. Genetic variation for survival and shell length of cultured red abalone (*Haliotis rufescents*) in Iceland. Journal of Shellfish Research, 18(2): 621–625.

Kanis, E., Refstie, T. and Gjedrem, T. 1976. A genetic analysis of egg, alevin and fry mortality in salmon (*Salmo salar*) sea trout (*Salmo trutta*) and rainbow trout (*Salmo gairdneri*). Aquaculture, 8: 259–268.

Karsi, A., Li P., Kim, S., Dunham, R. and Liu, Z.J. 2000. Performance traits-linked DNA markers and marker-assisted selection. In: Final Abstracts Guide. Plant and Animal Genome Conference VIII, January 9–12, 2000, San Diego, USA.

Kause, A., Ritola, O., Paananen, T., Mäntysaari, E. and Eskelinen, U. 2003. Selection against early maturity in large rainbow trout, *Oncorhynchus mykiss*: The quantitative genetics of sexual dimorphism and genotype-by-environment interactions. Aquaculture, 228: 53–68.

Kause, A., Ritola, O., Paananen, T., Mäntysaari, E. and Eskelinen, U. 2002. Coupling body weight and its composition: A quantitative genetic analysis in rainbow trout. Aquaculture, 211: 65–79.

Kause, A., Ritola, O., Paananen, T., Wahloos, H. and Mäntysaari, E. 2005. Genetic trends in growth, sexual maturity and skeletal deformations, and rate of inbreeding in a breeding programme for rainbow trout (*Onchorhynchus mykiss*). Aquaculture, 247: 177–187.

Kearsey, M. 1998. The principles of QTL analysis (a minimal mathematics approach). Journal of Experimental Botany, 49: 1619–1623.

Kharlamova, A.V. and Gulevitch, R.G. 1991. Peculiarities of emotional behaviour and adrenal function in American mink (*Mustela vison* Schreb.) during selection for behaviour. In: Trut, L.N., Osadchuk, L.V. and Borodin, P.M. (ed.). Evolutionary-Genetic and Genetic-Physiological Aspects of fur Animal Domestication. Novosibirsk: Institute of Cytology and Genetics, Large Sibirian Division of the USSR Academy of Sciences. (In Russian). pp. 219–231.

Kincaid, H.L., Bridges, W.R. and Limbach, B. 1977. Three generations of selection for growth rate in fall spawning rainbow trout. Transaction of the American Fisheries Society, 106: 621–629.

Kinghorn, B.P. 1981. Quantitative genetics in fish breeding. PhD. Thesis, University of Edinburgh, Edinburgh, 142pp.

Kinghorn, B.P. 1983. A review of quantitative genetics in fish breeding. Aquaculture, 31: 283–304.

Kirpichnikov, V.S. 1972. Methods and effectiveness of breeding the Ropshian carp. Communication in purpose of breeding, initial forms, and system of crosses. Soviet Genetics, 8: 996–1001.

Kittelsen, A., Venvik, T., Olesen, I., Fjæra, S.O. and Nervik, L. 2002. Management (Drift og røkt). GAN Forlag AS, 110pp.

Kjøglum, S., Henrion, M., Aasmundstad, T. and Korsgaard, I. 2008. Selective breeding can increase resistance of Atlantic salmon to furunculosis, infectious salmon anaemia and infectious pancreatic necrosis. Aquaculture Research, 39: 498–505.

Knapik, E.W., Goodman, A., Ekker, M., Chevrette, M., Delgado, J., Neuhauss, S., Shimoda, N., Driever, W., Fishman, M.C. and Jacob, H.J. 1998. A microsatellite genetic linkage map for zebrafish (*Danio rerio*). Nature Genetics, 18: 338–343.

Knibb, W.R., Gorshkova, G. and Gorshkov, S. 1997. Growth of strains of gilhead seabream, *Sparus aurata L.* Israeli Journal of Aquaculture Bamidgeh, 49: 57–66.

Kobayashi, S.-I., Alimuddin Morita, T., Miwa, M., Lu, J., Endo, M., Takeuchi, T. and Yoshizaki, G. 2007. Transgenic Nile tilapia (*Oreochromis niloticus*) over-expressing growth hormone show reduced ammonia excretion. Aquaculture, 270: 427–435.

Kocher, T.D., Lee, W.-J., Sobolewska, H., Penman, D. and McAndrew, B. 1998. A genetic linkage map of a cichlid fish, the tilapia (*Oreochromis niloticus*). Genetics, 148: 1225–1232.

Kocour, M., Manger, S., Rodina, M., Gela, D., Linhart, O. and Vandeputte, M. 2007. Heritability estimates for processing and quality traits in common carp (*Cyprinus carpio*, L.) using molecular pedigree. Aquaculture, 270: 43–50.

Kolstad, K., Grisdale-Helland, B. and Gjerde, B. 2004a. Family differences in feed efficiency in Atlantic salmon (*Salmo salar*). Aquaculture, 241: 169–177.

Kolstad, K., Heuch, P.A., Gjerde, B. and Gjedrem, T. 2004b. Genetic variation in resistance of Atlantic salmon (*Salmo salar*) to the salmon louse. Aquaculture, 247: 145–151.

Kolstad, K., Meuwissen, T.H.E. and Gjerde, B. 2005. Efficient design for doing genetic studies of feed efficiency in Atlantic salmon (*Salmo salar*). Aquaculture, 247: 153–158.

Kolstad, K., Ødegård, J., Tran, L., Nguyen, D. and Olesen, I. 2007. Improved methodology for analyzingsurvival data in fish breeding programs. Aquaculture, 272: 278–279.

Komen, J. 1990. Clones of common carp, *Cyprinus carpio*. New perspectives in fish research. Doctoral thesis, Agricultral University Wageningen, Wagening, The Netherlands, 169pp.

Kontali Analyse AS. Value creation in Norwegian Seafarming in 2004. (Verdiskaping i Norsk havbruksnæring i 2004). In: Akvakultur i Norge-2005. http://www.godfisk.no/page id=263&key = 1385.

Kosambi, D.D. 1944. The estimation of map distances from recombination values. Annals of Eugenics, 12: 172–175.

Lander, E.S. and Botstein, D. 1989. Mapping mendelian factors underlying quantitative traits using RFLP linkage maps. Genetics, 121: 185–199.

Langdon, C., Evans, F., Jacobson, D. and Blouin, M. 2003. Yields of cultured pacific oysters, *Crassostrea gigas* Thunberg, improved after one generation of selection. Aquaculture, 220: 227–244.

Langdon, C., Jacobson, D.P., Evans, F. and Blouin, M.S. 2000. The molluscan broodstock program. Improving pacific oyster broodstock through genetic selection. Journal of Shellfish Research, 19: 616.

Large, R.V. 1976. The influence of reproduction rate on the efficiency of meat production in animal populations. In: Lister, D., Rhodes, D.N., Fowler, V.R. and Fuller, M.F. (eds.). Meat Animals – Growth and Productivity. Plenum Press, New York and London, pp. 43–55.

Lee, B.Y., Lee, W.J., Streelman, J.T., Carleton, K.L., Howe, A.E., Hulata, G., Slettan, A., Stern, J.E., Terai, Y. and Kocher, T.D. 2005. A second-generation genetic linkage map of tilapia (*Oreochromis* spp.). Genetics, 170: 237–244.

Lee, W.J. 2003. Detection of QTL for salinity tolerance of tilapia. In: Final Abstracts Guide. Plant and Animal Genome Conference XI, January 11–15, San Diego, California.

Li, L. and Guo, X.M. 2004. AFLP-based genetic linkage maps of the Pacific oyster *Crassostrea gigas* Thunberg. Marine Biotechnology, 6: 26–36.

Liu, Z.J. and Cordes, J.F. 2004. DNA marker technologies and their applications in aquaculture genetics. Aquaculture, 238: 1–37.

Longalong, F.M., Eknath, A.E. and Bentsen, H.B. 1999. Response to bi-directional selection for frequency of early maturing females in Nile tilapia (*Oreochromis niloticus*). Aquaculture, 178: 13–25.

Lund, T., Gjedrem, T., Bentsen, H.B., Eide, D.M., Larsen, H.J.S. and Røed, K.H. 1995. Genetic variation in immune parameters and associations to survival in Atlantic salmon. Journal of Fish Biology, 46: 748–758.

Lush, J.L. 1949. Animal breeding plans. A book of the Iowa State College Press, Ames, IA, 443pp.

Lush, J.L. 1994. The genetics of populations. Published by Iowa Agriculture and Home Economics Experiment Station College of Agriculture, Iowa State University, Ames Iowa, 900pp.

Mahapatra, K.D., Gjerde, B., Saha, J.N., Reddy, P.V.G.K., Jana, R.K., Sahoo, M. and Rye, M. 2004. Realized genetic gain for growth in rohu (*Labeo rohita*). In manuscript.

Mair, G.C., Abucay, J.S., Beardmore, J.A. and Skibinski, D.O.F. 1995. Growth performance trials of genetically male tilapia (GMT) derived from YY males in *Oreochromis niloticus* L.: On-station comparisons with mixed sex and sex reversed male populations. Aquaculture, 137: 313–322.

Mair, G.C., Abucay, J.S., Skibinski, D.O.F., Abella, T.A. and Beardmore, J.A. 1997. Genetic manipulation of sex-ratio for the large-scale production of all-male tilapia, *Oreochromis niloticus*. Canadian Journal of Fisheries and Aquatic Sciences, 54: 396–404.

Malecot, G. 1948. Les mathematiques de l'heridite. Masson, Paris.

Maluwa, A.O. and Gjerde, B. 2006. Genetic evaluation of four strains of *Oreochromis shiranus* for harvest body weight in a diallel cross. Aquaculture, 259: 28–39.

Maluwa, A.O. and Gjerde, B. 2007. Response to selection for harvest body weight of *Oreochromis shiranus*. Aquaculture, 273: 33–41.

Maluwa, A.O., Gjerde, B. and Ponzoni, R.W. 2006. Genetic parameters and genotype by environment interaction for body weight of *Orochromis shiranus*. Aquaculture, 259: 47–55.

Mariasegaram, M. 2004. Using selective DNA pooling to map QTL affecting milk production traits in Australian dairy cattle. Ph.D Thesis. Melbourne University.

Marks, H.L. 1996. Long-term selection for body weight in Japanese quail under different environment. Poultry Science, 75: 1198–1203.

Massault, C., Bovenhuis, H., Haley, C. and de Koning, D.J. 2008. QTL mapping designs for aquaculture. Aquaculture, 285: 23–29.

McKay, L.R. and Gjerde, B. 1986. Genetic variation for a spinal deformity in Atlantic salmon, *Salmo salar*. Aquaculture, 52: 263–272.

Mendel, G.J. 1866. "Versuche uber Pflantzen – Hybriden". Verh. Naturforsch. Verein. Brunn, 4: 3–47.

Meuwissen, T.H.E. and Goddard, M. 1996. The use of marker haplotypes in animal breeding schemes. Genetics Selection and Evolution, 28: 161–176.

Meuwissen, T.H.E., Hayes, B.J. and Goddard, M.E. 2001. Prediction of total genetic value using genome-wide dense marker maps. Genetics, 157: 1819–1829.

Mitchell, G., Smith, C., Makower, M. and Bird, P.J.W.N. 1982. An economic appraisal of pig improvement in Great Brittain. 1. Genetic and production aspects. Animal Production, 35: 215–224.

Moav, R. and Finkel, A. 1975. Variability of intramuscular bones, vertebrate, ribs, dorsal fin rays and skeletal disorders in common carp. Theoretical and Applied Genetics, 46: 33–43.

Moav, R. and Wohlfarth, G.W. 1973. Carp breeding in Israel. In: R. Moav (ed.). Agricultural Genetics Selected topics. J. Wiley, New York, NY, 352pp.

Moav, R. and Wohlfarth. G.W. 1976. Two way selection for growth rate in common carp (*Cyprinus carpio* L.). Genetics, 82: 83–101.

Moav, R., Hulata, G. and Wohlfarth, G. 1975. Genetic differences between the Chinese and European races of the common carp. 1. Analysis of genotype – environment interactions. Heredity, 34: 323–340.

Moberg, G.P. 2000. Biological response to stress: Implicatios for animal welfare. In: Moberg, G.P. and Mench, J.A. (eds.). The biology of animal stress. Basic principles and implications for animal welfare. CABI Publishing, New York, 377pp.

Moen, T., Agresti, J.J., Cnaani, A., Moses, H., Famula, T.R., Hulata, G., Gall, G.A.E. and May, B. 2004a. A genome scan of a four-way tilapia cross supports the existence of a quantitative trait locus for cold tolerance on linkage group 23. Aquaculture Research, 35: 893–904.

Moen, T., Baranski, M., Sonesson, A. and Kjøglum, S. 2008a. Detection and fine-mapping of a major QTL for resistance to infectious pancreatic necrosis in Atlantic salmon (*Salmo salar*): population-level association between QTN genotype and trait. BMC Genomics, in press.

Moen, T., Fjalestad, K.T., Munck, H. and Gomez-Raya, L. 2004b. A multistage testing strategy for detection of quantitative trait loci affecting disease resistance in Atlantic salmon. Genetics, 167: 851–858.

Moen, T., Hayes, B., Baranski, M., Berg, P.R., Kjøglum, S., Koop, B.F., Davidson, W.S., Omholt, S.W. and Lien, S. 2008b. A linkage map of the Atlantic salmon (*Salmo salar*) based on EST-derived SNP markers. BMC Genomics, 9: 1–14.

Moen, T., Hoyheim, B., Munck, H. and Gomez-Raya, L. 2004c. A linkage map of Atlantic salmon (*Salmo salar*) reveals an uncommonly large difference in recombination rate between the sexes. Animal Genetics, 35: 81–92.

Moghadam, H., Poissant, J., Fotherby, H., Haidle, L., Ferguson, M. and Danzmann, R. 2007. Quantitative trait loci for body weight, condition factor and age at sexual maturation in Arctic charr (*Salvelinus alpinus*): comparative analysis with rainbow trout (*Oncorhynchus mykiss*) and Atlantic salmon (*Salmo salar*). Molecular Genetics and Genomics, 277: 647–661.

Moore, S.S., Whan, V., Davis, G.P., Byrne, K., Hetzel, D.J.S. and Preston, N. 1999. The development and application of genetic markers for the Kuruma prawn *Penaeus japonicus*. Aquaculture, 173: 1–4.

Mørkøre, T. and Einen, O. 2002. Relating sensory and instrumental texture analysis of Atlantic salmon. Doctor scientiarum thesis, Agricultural University of Norway, 27pp.

Mørkøre, T. and Rørvik, K.A. 2001. Seasonal variations in growth, feed utilisation and product quality of farmed Atlantic salmon (*Salmo salar*) transferred to seawater as 0+ smolts or 1+ smolts. Aquaculture, 199: 145–159.

Myers, J.M., Heggelund, P.O., Hudson, G. and Iwamoto, R.N. 2001. Genetics and broodstock management of coho salmon. Aquaculture, 197: 43–62.

Myers, J.M., Hershberger, W.K. and Iwamoto, R.N. 1986. The induction of tetraploidy in salmonids. Journal of the World Aquaculture Society, 17: 1–7.

Neely, K.G., Myers, J.M., Hard, J.J. and Shearer, K.D. 2008. Comparison of growth, feed intake, and nutrient efficiency in a selected strain of coho salmon (*Oncorhynchus kisutch*) and its source stock. Aquaculture, 283: 134–140.

Neira, R., Diaz, N.F., Gall, G.A.E., Gallardo, J.A., Lhorente, J.P. and Manterola, R. 2006a. Genetic improvement of coho salmon (*Oncorhynchus kisutsh*). I: Selection response and inbreeding depression on harvest weight. Aquaculture, 257: 9–17.

Neira, R., Diaz, N.F., Gall, G.A.E., Gallardo, J.A., Lhorente, J.P. and Alert, A. 2006b. Genetic improvement of coho salmon (*Oncorhynchus kisutch*). II: Selection response for early spawning date. Aquaculture, 257: 1–8.

Nell, J.A. and Hand, R.E. 2003. Evaluation of the progeny of second-generation Sydney rock oyster *Saccostrea glomerata* (Gold, 1850) breeding lines for resistance to QX disease *Marteilia sydneyi*. Aquaculture, 228: 27–35.

Nell, J.A., Smith, I.R. and Sheridan, A.K. 1999. Third generation evaluation of Sydney rock oyster *Saccostrea commercialis* (Iredale and roughly) breeding lines. Aquaculture, 170: 195–203.

Newkirk, G.F. and Haley, L.E. 1983. Selection for growth rate in the European oyster, *Ostrea edulis*: Response of second generation group. Aquaculture, 33: 149–155.

Nichols, K.M., Bartholomew, J. and Thorgaard, G.H. 2003a. Mapping multiple genetic loci associated with *Ceratomyxa shasta* resistance in *Oncorhynchus mykiss*. Disease of Aquatic Organisms, 56: 145–154.

Nichols, K.M., Young, W.P., Danzmann, R.G., Robison, B.D., Rexroad, C., Noakes, M., Phillips, R.B., Bentzen, P., Spies, I., Knudsen, K., Allendorf, F.W., Cunningham, B.M., Brunelli, J., Zhang, H., Ristow, S., Drew, R., Brown, K.H., Wheeler, P.A. and Thorgaard, G.H. 2003b. A consolidated linkage map for rainbow trout (*Oncorhynchus mykiss*). Animal Genetics, 34: 102–115.

Nilsson, J. 1992. Genetic variation in resistance of *Arctic char* to fungal infection. Journal of Aquatic Animal Health, 4: 36–47.

Notter, D.R. 1999. The importance of genetic population diversity in livestock populations of the future. Journal of Animal Science, 77: 61–69.

Ødegård, J., Olesen, I., Gjerde, B. and Klemetsdal, G. 2006. Evaluation of statistical models for genetic analysis of challenge test data on furunculosis resistance in Atlantic salmon (*Salmo salar*): Prediction of field survival. Aquaculture, 259: 116–123.

Ødegård, J., Olesen, I., Gjerde, B. and Klemetsdal, G. 2007. Positive genetic correlation between resistance to bacterial (furunculosis) and viral (infectious salmon anaemia) diseases in farmed Atlantic salmon (*Salmo salar*). Aquaculture, 273–177.

Ødegård, J., Yazdi, M.H., Sonesson, A.K. and Meuwissen, T.H.E. 2009. Incorporating desirable genetic characteristics from an inferior into a superior population using genomic selection. Genetics, 181: 737–745.

Ohara, E., Nishimura, T., Nagakura, Y., Sakamoto, T., Mushiake, K. and Okamoto, N. 2005. Genetic linkage maps of two yellowtails (*Seriola quinqueradiata* and *Seriola lalandi*). Aquaculture, 244: 41–48.

Okamoto, N., Tayaman, T., Kawanobe, M., Fujiki, N., Yasuda, Y. and Sano, T. 1993. Resistance of a rainbow trout strain to infectious necrosis. Aquaculture, 117: 71–76.

Olesen, I., Groen, A.F. and Gjerde, B. 2000. Definition of animal breeding goals for sustainable production systems. Journal of Animal Science, 78: 570–582.

Oppedal, F., Taranger, G.L. and Hansen, T. 2003. Growth performance and sexual maturation in diploid and triploid Atlantic salmon (*Salmo salar* L.) in seawater tanks exposed to continuous light or simulated natural photoperiod. Aquaculture, 215: 145–162.

Ozaki, A., Sakamoto, T., Khoo, S.K., Nakamura, K., Coimbra, M.R.M., Akutsu, T. and Okamoto, N. 2001. Quantitative trait loci (QTLs) associated with resistance/susceptibility to infectious pancreatic necrosis virus (IPNV) in rainbow trout (*Oncorhynchus mykiss*). Molecular Genetics and Genomics, 265: 23–31.

Pechsiri, J. and Yakupitiyage, A. 2005. A comparative study of growth and feed utilization efficiency of sex-reversed diploid and triploid Nile tilapia, *Oreochromis niloticus* L. Aquaculture Research, 36: 45–51.

Perry, G.M.L., Ferguson, M.M., Sakamoto, T. and Danzmann, R.G. 2005. Sex-linked quantitative trait loci for thermotolerance and length in the rainbow trout. Journal of Heredity, 96: 97–107.

Pierce, L.R., Palti, Y., Silverstein, J.T., Barrows, F.T., Hallerman, E.M. and Parsons, J.E. 2008. Family growth response to fish meal and plant-based diets shows genotype x diet interaction in rainbow trout (*Onchorhynchus mykiss*). Aquaculture, 278: 37–42.

Ponzoni, R.W., Hamzah, A.b., Saadiah, S.T. and Kamaruzzaman, N. 2003. Phenotypic and genetic parameters for live weight in two environments in a selected line of Nile tilapia in Malaysia. IAGA, Genetics in Aquaculture VIII. 9–15 November 2003, Puerto Varas, Chile: 83.

Ponzoni, R.W., Nguyen, N.H. and Khaw, H.L. 2007. Investment appraisal of genetic improvement programs in Nile tilapia (*Oreochromis niloticus*). Aquaculture, 269:187–199.

Poompuang, S. and Hallerman, E.M. 1997. Toward detection of quantitative trait loci and marker-assisted selection in fish. Reviews in Fisheries Science, 5: 253–277.

Potteringer, T.G. 2006. Context dependent differences in growth of two rainbow trout (*Oncorhynchus mykiss*) lines selected for divergent stress responsiveness. Aquaculture, 256: 140–147.

Price, E.O. 1984. Behavioural aspects of animal domestication. Quarterly Review of Biology, 59: 1–32.

Price, E.O. 2002. Animal Domestication and Behaviour. CABI publishing, New York, 297pp.

Purdom, C.E. 1972. Induced polyploidy in place (*Pleuronectes platessa*) and its hybrid with flounder (*Platichthys flesus*). Heredity, 29: 11–24.

Purdom, C.E. 1993. Genetics and Fish Breeding. Chapman and Hall. London-Glasgow-New York-Tokyo-Melbourne-Madras, 277pp.

Quillet, E., Guillou, S.L., Aubin, J. and Fauconneau, B. 2005. Two-way selection for muscle lipid content in pan-size rainbow trout (*Oncorhynchus mykiss*). Aquaculture, 245: 49–61.

Rahman, M.A. and Maclean, N. 1999. Growth performance of transgenic tilapia containing an exogenous piscine growth hormone gene. Aquaculture, 173: 333–346.

Rauw, W.M., Kanis, E., Noorhuixen-Stassen, E.N. and Grommers, F.J. 1998. Undisirable side effects of selection for high production efficiency in farm animals: A review. Livestock Production Science, 56: 15–33.

Reddy, P.V.G.K., Gjerde, B., Tripathi, S.D., Jana, R.K., Mahapatra, K.D., Gupta, S.D., Saha, J.N., Sahoo, M., Lenka, S., Govindassamy, P., Rye, M. and Gjedrem, T. 2002. Growth and survival of six stocks of rohu (*Labeo rohita*) in mono and polyculture production systems. Aquaculture, 203: 239–250.

Reddy, P.V.G.K., Jana, R.K., Mahapatra, K.D., Saha, J.N., Meher, P.K., Sahoo, M., Mohanty, J., Sarkar, S., Gjerde, B. and Rye, M. 2003. Selective breeding of rohu, May 1992 – March 1996 and Genetic improvement of rohu for growth through selective breeding, April 1997 – June 2003. Final report on The Indo-Norwegian Collaboration Project, CIFA and AKVAFORSK, 56pp.

Refstie, T. 1983a. Hybrids between salmonid species. Growth rate and survival in seawater. Aquaculture, 33: 281–285.

Refstie, T. 1983b. Induction of diploid gynogenesis in Atlantic salmon and rainbow trout using irradiated sperm and heat shock. Canadian Journal of Zoology, 61: 2411–2416.

Refstie, T. and Gjedrem, T. 1975. Hybrids between salmonidae species, hatchability and growth rate in the freshwater period. Aquaculture, 6: 333–342.

Refstie, T. and Steine, T. 1978. Selection experiments with salmon III. Genetic and environmental sources of variation in length and weight of Atlantic salmon in the freshwater phase. Aquaculture, 14: 221–234.

Refstie, T., Mørkøre, T., Johansen, H. and Gjøen, H.M. 1999. Better texture in Norwegian salmon through selective breeding. (Bedre texture hos norsk laks gjennom avlsarbeidet). AKVAFORSK – Report 12/99, 19pp.

Reid, D.P., Szanto, A., Glebe, B., Danzmann, R.G. and Ferguson, M.M. 2004. QTL for body weight and condition factor in Atlantic salmon (*Salmo salar*): comparative analysis with rainbow trout (*Oncorhynchus mykiss*) and Arctic charr (*Salvelinus alpinus*). Heredity, 94: 166–172.

Riquet, J., Coppieters, W., Cambisano, N., Arranz, J.-J., Berzi, P., Davis, S.K., Grisart, B., Farnir, F., Karim, L., Mni, M., Simon, P., Taylor, J.F., Vanmanshoven, P., Wagenaar, D., Womack, J.E. and Georges, M. 1999. Fine-mapping of quantitative trait loci by identity by descent in outbred populations: Application to milk production in dairy cattle. Proceeding of the. National Academy Sciences United States of America, 96: 9252–9257.

Roberts, R.C. 1966. The limits to artificial selection for body weight in the mouse. I. The limits attained in earlier experiments. Genetic Research, 8: 361–375.

Robertson, A. 1955. Selection response and the properties of genetic variation. Cold Spring Harbor Symposia on Quantitative Biology, 20: 166–177.

Robinson, N. and Hayes, B. 2008. Modelling the use of gene expression profiles with selective breeding for improved disease resistance in Atlantic salmon (*Salmo salar*). Aquaculture, 285: 38–46.

Robinson, N., Goddard, M. and Hayes, B. 2008. Use of gene expression data for predicting continuous phenotypes for animal production and breeding. Animal, 2: 1413–1420.

Robison, B.D., Wheeler, P.A., Sundin, K., Sikka P. and Thorgaard, G.H. 2001. Composite interval mapping reveals a major locus influencing embryonic development rate in rainbow trout (*Oncorhynchus mykiss*). Journal of Heredity, 92: 16–22.

Robison, O.W. and Luempert, L.G. 1984. Genetic variation in weight and survival of brook trout (*Salvelinius fontinalis*). Aquaculture, 38: 155–170.

Rocha, E., Matic, I. and Taddei, F. 2002. Over-representation of repeats in stress response genes: A strategy to increase versatility under stressful conditions? Nucleic Acids Research, 30: 1886–1894.

Rodriguez, M.F., LaPatra, S., Williams, S., Famula, T. and May, B. 2004. Genetic markers associated with resistance to infectious hematopoietic necrosis in rainbow and steelhead trout (*Oncorhynchus mykiss*) backcrosses. Aquaculture, 241: 93–115.

Rørå, A.M.B., Mørkøre, T. and Einen, O. 2001. Primary processing (Evisceration and filleting). In: Kestin, S.C. and Warriss, P.D. (eds.). Farmed Fish Quality. Fishing Newes Books, 430pp.

Rosendal, G.K., Olesen, I., Bentsen, H.B., Walløe Tvedt, M. and Bryde, M. 2005. Strategies and regulations pertaining to access to and legal protection of aquaculture genetic resources. FNI Report 7/2005.

Rosseland, B.O., Skogheim, O.K. and Sevaldrud, I.H. 1986. Acid deposition and effects on Nordic Europe. Damage to fish populations in Scandinavia continue apace. Water, Air, and Soil Pollution, 130: 899–904.

Rothschild, M.F., Larson, R., Jacobson, C. and Pearson, P. 1991. PvuII polymorphisms at the porcine oestrogen receptorlocus (ESR). Animal Genetics, 22: 448.

Ruiz-Verdugo, C.A., Ramirez, J.L., Allen, S.K. and Ibarra, A.M. 2000. Triploid catarina scallop (*Argopecten ventricosus* Sowerby II, 1842): Growth, gametogenesis, and suppression of functional hermafroditism. Aquaculture, 186: 13–32.

Ruzzante, D.E. 1994. Domestication effects on aggressive and schooling behaviour in fish. Aquaculture, 120: 1–24.

Rye, M and Gjedrem, T. 2005. Measuring genetic change. In: Gjedrem, T (ed.). Selection and Breeding Programs in Aquaculture. Springer, Berlin, Heidelberg, 364pp.

Rye, M. 1991. Prediction of carcass composition in Atlantic salmon by computerised tomography. Aquaculture, 99: 35–48.

Rye, M. and Gjerde, B. 1996. Phenotypic and genetic parameters of composition traits and flesh colour in Atlantic salmon. Aquaculture Research, 27: 121–133.

Rye, M. and Mao, I.L. 1998. Nonadditive genetic effects and inbreeding depression for body weight in Atlantic salmon (*Salmo salar* L.). Livestock Production Science, 57: 15–22.

Rye, M. and Refstie, T. 1995. Phenotypic and genetic parameters of body size traits in Atlantic salmon, *Salmo salar* L. Aquaculture Research, 26: 875–885.

Rye, M., Lillevik, K.M. and Gjerde, B. 1990. Survival in early life of Atlantic salmon and rainbow trout: estimates of heritabilities and genetic correlations. Aquaculture, 89: 209–216.

Rye, M., Storebakken, T., Gjerde, B. and Ulla, O. 1994. Efficient selection for colour in Atlantic salmon (Effektiv seleksjon for innfarging hos laks). Norsk Fiskeoppdrett, 11A: 22–24.

Sakamoto, T., Danzmann, R.G., Gharbi, K., Howard, P., Ozaki, A., Khoo, S.K., Woram, R.A., Okamoto, N., Ferguson, M.M., Holm, L.-E., Guyomard, R. and Hoyheim, B. 2000. A microsatellite linkage map of rainbow trout (*Oncorhynchus mykiss*) characterized by large sex-specific differences in recombination rates. Genetics, 155: 1331–1345.

Sakamoto, T., Danzmann, R.G., Okamoto, N., Ferguson, M.M. and Ihssen, P.E. 1999. Linkage analysis of quantitative trait loci associated with spawning time in rainbow trout (*Oncorhynchus mykiss*). Aquaculture, 173: 1–4.

Salte, R., Bentsen, H.B., Moen, T., Tripathy, S., Omholt, S., Bakke, T.A., Ødegård, J. and Hansen, L.P. 2009. Prospects for a genetic management strategy to control *Gyrodactylus salaris* infection in wild Atlantic salmon (*Salmo salar*) stocks. Canadian Journal of Fisheries and Aquatic Sciences, submitted.

Sather, A.P. 1987. A note on the changes in leg weakness in pigs after being transferred from confinement housing to pasture lots. Animal Production, 44: 450–453.

Schaperclaus, W. 1962. Trate de pisciculture en Etang. Vigot Freres, Paris, pp. 208–227.

Schlötterer, C. 2000. Evolutionary dynamics of microsatellite DNA. Chromosoma, 109: 365–371.

Segenbusch, R. and Meske, C. 1967. Auf dem Wege zum grætenlosen Karpfen. Der Züchter, 37: 271–274.

Selvamani, M.J.P., Degnan, S.M. and Degnan, B.M. 2001. Microsatellite genotyping of individual abalone larvae: Parentage assignment in aquaculture. Marine Biotechnology, 3: 478–485.

Sharp, L.W. 1934. Introduction to cytology. McGrav-Hill, New York.

Siitonen, L. and Gall, G.A.E. 1989. Response to selection for early spawn date in rainbow trout, *Salmo gairdneri*. Aquaculture, 78: 153–161.

Skaala, Ø., Høyheim, B., Glover, K. and Dahle, G. 2004. Microsatellite analysis in domesticated and wild Atlantic salmon *(Salmo salar* L.): Allelic diversity and identification of individuals. Aquaculture, 240: 131–143.

Skagemo, V., Sonesson, A.K., Meuwissen, T. and Rye, M. 2008. Increased profits in aquaculture by optimise the selection of parents from the nucleus to the grow-out producers. Aquaculture, submitted.

Slettan, A. 2004. Genetic tracability system. Gen-Treck. Concluding report for user-driven R&D project. Research Council of Norway.

Smith, H. 1936. A discriminant function for plant selection. Annals of Engineering, 7: 240–250.

Smith. P.J. and Conroy. A.M. 1992. Loss of genetic variation in hatchery-produced abalone, *Haliotis iris*. N. Z. Journal of Marine and Freshwater Research, 26: 81–85.

Sonesson. A.K. 2007. Within-family marker-assisted selection for aquaculture species. Genetics Selection Evolution, 39: 301–317.

Spelman, R.J., Garrick, D.J. and van Arendonk, J.A.M. 1999. Utilisation of genetic variation by marker assisted selection in commercial dairy cattle populations. Livestock Production Scicience, 59: 51–60.

Staelens, J., Rombaut, D., Vercauteren, I., Argue, B., Benzie, J. and Vuylsteke, M. 2008. High-Density Linkage Maps and Sex-Linked Markers for the Black Tiger Shrimp (*Penaeus monodon*). Genetics, 179: 917–925.

Standal, M. and Gjerde, B. 1987. Genetic variation in survival of Atlantic salmon during the sea-rearing period. Aquaculture, 66: 197–207.

Storset, A., Strand, C., Wetten, M., Kjøglum, S. and Ramstad, A. 2007. Response to selection for resistance against infectious pancreatic necrosis in Atlantic salmon (*Salom salar* L.). Aquaculture, 272: 62–68.

Suarez, J.A., Gitterle, T., Angarita, M.R. and Rye, M. 1999. Genetic parameters for harvest weight and pond survival in *Litopenaeus vannamei*. European Aquaculture Society, Special Publication. No 27, June 1999: 232–233.

Sun, X. and Liang, L. 2004. A genetic linkage map of common carp (*Cyprinus carpio* L.) and mapping of a locus associated with cold tolerance. Aquaculture, 238: 165–172.

Sundin, K., Brown, K.H., Drew, R.E., Nichols, K.M., Wheeler, P.A. and Thorgaard, G.H. 2005. Genetic analysis of a development rate QTL in backcrosses of clonal rainbow trout, *Oncorhynchus mykiss*. Aquaculture, 247: 75.

Swan, A.A., Thompson, P.A. and Ward, R.D. 2007. Genotype x environment interactions for weight in Pacific oysters (*Crassostrea gigas*) on five Australian farms. Aquaculture, 265: 91–101.

Sylven, S., Rye, M. and Simianer, H. 1991. Interaction of genotype with production system for slaughter weight in rainbow trout (*Oncorhynchus mykiss*). Livestock Production Science, 28: 253–263.

Tautz, D. 1989. Hypervariability of simple sequences as a general source for polymorphic DNA markers. Nucleic Acids Research, 17: 6463–6471.

Taylor, R.S., Wynne, J.W., Kube, P.D. and Elliott, N.G. 2007. Genetic variation of resistance to amoebic gill disease in Atlantic salmon (*Salmo salar*) assessed in a challenge system. Aquaculture, 272: 94–99.

Teichert-Coddington, D. 1983. Divergent selection for prematuration body weight in tilapia niloticus. M.S. Thesis, Auburn University, Auburn, AL.

Teichert-Coddington, D.R. and Smitterman, R.O. 1988. Lack of response by tilapia nilotica to mass selection for rapid early growth. Transaction of the American Fisheries Society, 117: 297–300.

Thodesen, J., Gjerde, B., Grisdale-Helland, B. and Storebakken, T. 2001. Genetic variation in feed intake, growth and feed utilization in Atlantic salmon (*Salmo salar*). Aquaculture, 194: 273–281.

Thodesen, J., Grisdale-Helland, B., Helland, S.J. and Gjerde, B. 1999. Feed intake, growth and feed utilization of offspring from wild and selected Atlantic salmon (*Salmo salar*). Aquaculture, 180: 237–246.

Thorgaard, G.H., Scheerer, P.D., Hershberger, W.K. and Myers, J.M. 1990. Androgenetic rainbow trout produced using sperm from tetraploid males show improved survival. Aquaculture, 85: 215–221.

Tong, J. and Chu, K.H. 2002. Genome mapping in aquatic animals: Progress and future perspectives. Russian Journal of Genetics, 38: 612–621.

Uriwan, S. and Doyle, R.W. 1986. Replicate variance and the choice of selection procedures for tilapia (*Oreochromis niloticus*) stock improvement in Thailand. Aquaculture, 57: 93–98.

Van Vleck, L.D., Pollak, E.J. and Oltenacu, E.A.B. 1987. Genetics for the Animal Sciences. W.H. Freeman and Company, New York, 391pp.

Vandeputte, M. and Prunet, P. 2002. Genetics of adaptation in fish: Effects of domestication, stress resistance and adaptation to the environment. Productions Animals, 15 (5): 365–371.

Vangen, O. 1984. Future breeding program in pigs in a situation with artificial insemination. (Framtidig avlsopplegg på svin i en KS-situasjon). Aktuelt fra Statens fagtjeneste for landbruket, 1: 300–306.

Venter, J.C., Adams, M.D., Myers, E.W., Li, P.W., Mural, R.J., Sutton, G.G., Smith, H.O., Yandell, M., Evans, C.A., Holt, R.A., Gocayne, J.D., et al. 2001. The sequence of the human genome. Science, 291: 1304–1351.

Villanueva, B., Veerspoor, E. and Visscher, P.M. 2002. Parental assignment in fish using microsatellite genetic markers with finite numbers of parents and offspring. Animal Genetics, 33: 33–41.

Vos, P., Hogers, R., Bleeker, M., Reijans, M., Lee, T., Hornes, M., Frijters, A., Pot, J., Peleman, J., Kuiper, M. and Zabeau, M. 1995. AFLP: A new technique for DNA fingerprinting. Nucleic Acids Research, 23: 4407–4414.

Waldbieser, G.C., Bosworth, B.G., Nonneman, D.J. and Wolters, W.R. 2001. A microsatellite-based genetic linkage map for channel catfish, *Ictalurus punctatus*. Genetics, 158: 727–734.

Wang, L., Song, L., Chang, Y., Xu, W., Ni, D. and Guo, X. 2005. A preliminary genetic map of Zhikong scallop (*Chlamys farreri* Jones et Preston 1904). Aquaculture Research, 36: 643–653.

Waterston, R., Lindblad-Toh, K., Birney, E., Rogers, J., Abril, J.F., Agarwal, P., Agarwala, R., Ainscough, R., Alexandersson, M., An, P., et al. 2002. Initial sequencing and comparative analysis of the mouse genome. Nature, 420: 520–562.

Watson, J. and Crick, F.H.C. 1953. Molecular structure of nucleic asids. Nature, 171: 737–738.

Wilson, K., Li, Y., Whan, V., Lehnert, S., Byrne, K., Moore, S., Pongsomboon, S., Tassanakajon, A., Rosenberg, G. and Ballment, E. 2002. Genetic mapping of the black tiger shrimp *Penaeus monodon* with amplified fragment length polymorphism. Aquaculture, 204: 297–309.

Wohlfarth, G.W. 1993. Heterosis for growth rate in common carp. Aquaculture, 113: 31–46.

Wohlfarth, G.W., Moav, R. and Hulata, G. 1983. A genotype – environment interaction for growth rate in common carp, growing in intensive manured ponds. Aquaculture, 33: 187–195.

Wright, S. 1921. Systems of mating. Genetics, 6: 111–178.

Xiang, J., Li, F., Zhang, C., Zhang, X., Yu, K., Zhou, L. and Wu, C. 2006. Evaluation of induced triploid shrimp *Penaeus (Fenneropenaeus) chinensis* cultured under laboratory conditions. Aquaculture, 259: 108–115.

Yu, Z.N. and Guo. X.M. 2003. Genetic linkage map of the eastern oyster *Crassostrea virginica* Gmelin. Biological Bulletin, 204: 327–338.

Zbikowska, H.M. 2003. Fish can be first-advances in fish transgenesis for commercial applications. Transgenic Research, 12: 379–389.

Index

Note: Page numbers followed by 'f' and 't' refer to figures and tables respectively.

A

Accuracy, 83, 85, 95, 96, 97, 98, 100, 102, 113, 114, 128, 158
Additive, 16, 29, 36, 37, 51, 52, 58, 60, 62, 85, 87, 90, 98, 108, 109, 110, 111, 114, 116, 153, 166, 172
  effect, 36, 52, 85, 93, 187
Allele, 26, 29, 30, 31f, 32, 36, 45, 46, 47, 49, 52, 54, 58, 62, 65, 90, 91, 93–94, 98, 113, 126, 152f, 156, 157f, 158, 159f, 162, 164, 187
  frequency, 58, 93, 187
Amino acid, 25, 26, 29, 187
Ancestor, 43, 44, 48, 66, 66f, 95, 115, 117
  common, 43–44
Androgenesis, 46, 171, 187
Aquaculture, 1–3, 5–11, 13–23, 28, 30, 31, 34, 36, 38t, 56, 63, 64, 66, 70, 74, 75, 78, 80, 81, 87, 88, 89f, 96, 99, 105, 144, 151, 153, 154t, 155, 156, 158, 160, 162, 166, 181, 182, 187

B

Base population, 17, 18f, 19, 45, 48, 59, 62, 65–66, 107f, 108, 126, 144, 187
BKD (bacterial kidney disease), 41, 42f, 181, 187
BLUP (Best Linear Unbiased Prediction), 102, 114, 117, 187
Breed, 3, 48, 87, 172, 187
Breeding
  cross, 50–51, 87, 88–90, 91, 108, 144, 183, 189
  goal, 3, 7, 14, 56, 67–80, 82, 84, 87, 99, 100, 101, 102, 105, 113, 115, 116, 119, 124, 125, 128, 136, 139, 142, 143, 144, 145, 146, 147, 148, 149, 161, 176, 188
  program, 3, 8f, 9, 10–11, 13, 14, 15, 20, 23, 36, 38, 45, 48, 52, 57, 59, 63–85, 87, 88, 94, 96, 98, 101, 102, 105, 108, 110, 113, 114, 115, 119, 123, 124, 125, 129, 131–144, 145–149, 151–167, 175–183
  pure, 3, 52, 87, 90–91, 194
  station, 11, 124, 132, 136, 137, 138, 139, 140, 142, 147, 188
  strategies, 3, 36, 87–91, 163, 188
Breeding value
  prediction, 37, 38, 97, 113, 114, 115, 146
  true, 91
Broodstock, 6, 9, 10, 21, 48, 64, 65, 80, 82, 84, 91, 93, 96, 98, 102, 105, 106, 107, 108, 109, 111, 113, 114, 126, 132, 133f, 136, 137, 138, 140, 141, 142, 143, 148, 160, 163, 164, 174, 175–176, 179, 188

C

Cell
  diploid, 169, 189
  haploid, 169, 191
Centromere, 169, 188
Challenge test, 14, 15f, 20, 72, 72f, 73, 73f, 78, 81, 82, 97, 137, 138, 188, 714
Chromosome, 26, 27f, 28, 29, 33, 153, 156, 158, 162, 163, 165, 166, 169, 170, 171, 173, 174, 188
Coefficient
  inbreeding, 45, 46, 48, 49, 49t, 148, 170, 191
  regression, 43, 84, 116
  variation, 22, 35, 68, 179, 180t
Competition, 2, 7, 142
Connectedness, 110, 128, 188
Correlated response, 19, 57–58, 69, 99–100, 176, 181, 188

Correlation
covariance, 35, 36, 40
environmental, 41, 41f, 82, 116, 189
genetic, 20, 40, 41, 42t, 57, 67, 69, 71, 71t, 72, 73, 74, 82, 99, 100, 102, 113, 115, 120, 121t, 123, 145, 146, 165, 190
phenotypic, 41, 41f, 64, 143, 194
Crossbreeding, 50–51, 87, 88–90, 91, 108, 144, 183, 189
Crosses
diallel, 19, 50, 51, 66
heterosis, 90
Cryopreservation, 128

D
Diallel, 19, 50, 51, 66
Dissemination, 8, 99, 106, 139, 140–143, 144, 149
Distribution
normal, 32, 179, 193
skewed, 179
DNA, 25, 26, 27, 28, 40, 106, 111, 151, 152, 153, 156, 159, 159f, 160, 163, 164, 166, 167, 169, 176, 189
markers, 40, 111, 151–152, 156, 159, 160, 163, 164, 176
Domestication, 5–11, 63, 78, 81, 143, 179, 180, 181, 189
Dominance
deviation, 98
variance, 36, 51, 87, 109

E
Effective population size, 47, 59, 60, 62, 65, 91, 94, 105, 106, 126, 164, 189
Environment, 5, 6, 7, 34, 36, 41, 52, 53, 63, 94, 97, 105, 109, 110, 114, 119–124, 145, 148, 149, 176
common, 97, 109
Enzymes, 189
Epistasis, 29, 50, 87, 190

F
Family
full-sib, 33, 40, 41, 42f, 72f, 73, 97, 102, 108, 109, 114–115, 123, 140, 162
half-sibs, 36, 108, 115, 136
FCE (feed conversion efficiency), 69–70, 190
FCR (feed conversion ratio), 14, 57, 58, 69–70, 82, 99, 190
Fecundity, 6, 9, 10, 22, 23, 45, 48, 58, 59, 64, 78, 93, 94, 140, 141, 144, 148, 154, 176, 190
Fertility, 49, 59, 81, 98, 146, 178, 190
Fitness, 29, 48, 49, 50, 53, 68, 87, 99, 100, 145, 146, 190

G
Gametes
female, 9, 25, 27f, 40, 97, 169, 170, 171, 193
male, 11, 25, 26, 27f, 106, 134, 140, 166, 171, 173, 174, 196
Gametogenesis, 190
GAS (gene-assisted selection), 161, 162, 190
Gene
dominance, 29, 36, 51, 87, 98, 109
frequency, 190
recessive, 29, 48, 94
Generation
interval, 8, 8f, 17, 55, 59, 65, 99, 178, 190
length, 55
Genetic
covariance, 36, 40
drift, 47, 59, 94, 164, 191
gain, 13, 14t, 15, 17, 20, 21, 22, 23, 48, 54, 55, 56t, 62, 64, 67, 75, 82, 88, 91, 96, 99, 102, 110, 116, 119, 124, 125, 127f, 128, 129, 134f, 140–143, 163, 167, 175, 177, 178, 191
marker, 40, 97, 116, 138, 150, 155, 156, 163–164, 183, 191
molecular, 151, 155, 161f, 193
parameters, 36, 58, 64, 101, 115, 116, 131, 143, 144
variance, 36, 37, 48, 50, 58, 87, 90, 98, 108, 109, 111, 148, 156, 160, 162, 191
Genotype, 30, 31f, 32, 34, 35, 36, 38, 40, 41, 52, 57, 58, 113, 114, 116, 119–124, 138, 148, 152f, 157, 159, 160, 161, 162, 163, 166, 191
frequency, 30
Genotype–environment
interaction, 119–124, 148, 191
Genotypic value, 116
GIFT (Genetic Improvement of Farmed Tilapias), 17, 18f, 66, 121
Gonad, 74, 146, 171, 172, 191
Gynogenesis, 46, 169–171, 191

H
Heredity, 29, 191
Heritability,
broad sense, 37, 73
narrow sense, 37
realized, 37
Hermaphrodite, 191
Heterosis, 50–51, 51t, 88, 89f, 90, 91, 144, 191

Heterozygote, 30, 49, 88, 152f, 154
Homozygote, 45, 191
Hybrid vigour, 3, 50, 88, 90, 171

I
Inbreeding
- coefficient, 45, 46, 48, 49, 148
- depression, 10, 48–49, 192
- line, 48, 50
- rate of, 47, 49, 105, 108, 111, 133f, 134f, 163

Interaction, 25, 29, 36, 50, 90, 119–124, 125, 126, 131, 138, 144, 148, 160, 164

L
LDLA, 158, 192
LD-MAS, 162, 192
LE-MAS, 160, 161, 162, 192
Lipid, 16, 192
Locus/loci, 29, 30, 31f, 36, 45, 50, 54, 58, 60, 62, 90, 153, 155–159, 161, 163, 192

M
MAS (marker assisted selection), 52, 155, 160, 161f, 162, 192
Mass spawning, 105–106, 111, 163, 192
Maternal effect, 108, 109, 115
Mating design
- factorial, 109, 111, 190
- full-sibs, 45, 46t, 105
- half-sibs, 45, 46t, 105
- nested, 100, 108–109, 111, 193
- repeated, 99, 110, 128, 129, 195
- single pairs, 106–108, 111

Maturation, 15, 17, 18, 23, 74, 80, 82, 97, 99, 115, 131, 137, 139, 147, 155t, 172, 177
Meiosis, 26, 27f, 169, 170, 192
Mendel's laws, 26
Microsatellite, 97, 138, 151, 152, 152f, 153, 159f, 163, 164, 192
Migration, 58, 59, 93, 192
Mitosis, 26, 193
Molluscs, 1, 136, 153, 165, 172, 193
Monoculture, 18, 19f, 89f, 121, 122f, 193
Multiplier station, 11, 142
Mutation, 29, 30, 58, 59, 62, 94, 153, 156, 158, 161, 162, 193

N
Ne, 17, 47, 52, 60, 126, 148, 189, 193
Non additive
- effect, 193
- variance, 36

Nucleus
- breeding unit, 193
- cell, 28

O
Omnivorous, 1, 193
Oogenese, 27f
Ovary, 193
Ovulation, 80, 94, 193

P
Parameters, 35, 36, 37, 55, 58, 64, 71, 82, 100, 101, 115, 116, 119, 125, 131, 143, 144
Path, 40, 41f, 44, 57, 140
Pedigree, 3, 11, 36, 59, 65, 87, 91, 94, 95, 95t, 96, 97, 105, 113, 129, 160, 164, 183, 193
Phenotype, 11, 13, 29, 30, 34, 35, 37, 40, 41f, 52, 53, 54, 57, 75, 91, 101t, 114, 125, 155, 156, 161, 165, 166, 194
PIT tag, 136, 194
Pleiotropy, 40, 194
Polar body, 169, 170, 171, 173, 194
Polyculture, 18, 19f, 75, 89f, 121, 122f, 194
Population
- base, 17, 18, 19, 45, 48, 59, 62, 65–66, 107f, 108, 126, 144, 187
- closed, 45, 47, 52, 87
- control, 60, 110, 126, 129
- synthetic, 65, 66f
- wild, 53, 65, 94, 164

Progeny testing, 59, 94, 95t, 98–99, 144, 174, 194
Protein, 1, 14, 16, 23, 26, 29, 32, 60, 61f, 68, 70, 124, 165, 194
Purebreeding, 3, 52, 87, 90–91, 194

Q
QTL (Quantitative Trait Loci), 155–159, 160, 162, 194
Qualitative trait, 30, 194
Quality
- colour, 59, 76, 87
- fat%, 38t, 76
- fat distribution, 76
- filet%, 38t
- texture, 38t, 76–77

Quantitative
- genetics, 3, 31, 37, 131, 143, 144, 169, 194
- trait, 30–32, 33, 37, 38, 40, 58, 60, 156, 160, 194
- variation, 155

R
Random
drift, 126
mating, 65, 88
sample, 128, 132, 138, 194
Recapture frequency, 79, 80, 194
Recessive allele, 194
Regression, 42–43, 84, 115, 116, 165, 195
coefficient, 43, 116
Relationship, 40, 43f, 43–44, 45, 46t, 58, 70, 74, 83, 90, 96, 98, 114, 115, 117, 129, 138, 145, 195
Relatives, 38, 41, 43, 44f, 46t, 48, 52, 53, 59, 91, 96, 100, 113, 132, 138, 162
Repeatability, 152
Retention
energy, 14t, 14, 189
protein, 14t, 194
RNA, 189

S
Sample, 106, 126, 128, 132, 136, 138, 159, 164, 165, 195
Sea ranching, 15, 79, 80, 195
Selection
accuracy of, 96
artificial, 5, 53–54, 62, 63, 146, 187
BLUP, 117
broodstock, 9, 21, 138, 140, 141
combined, 90, 98, 99, 100–101, 103, 144, 188
correlated response, 19, 57–58, 69, 99–100, 176, 181, 188
differential, 14, 16f, 20, 54, 55, 195
directional, 5, 14, 19, 53, 54
disease resistance, 23, 59, 70–74, 81, 97, 99, 100, 115, 139, 146, 147, 155, 160, 163, 165, 177
family, 17, 59, 79, 83, 95, 96–98, 100, 102, 103f, 103, 105, 114, 132, 141f, 144, 160, 190
within family, 17, 83, 98, 100, 102, 103, 105, 144, 160, 197
full-sib, 96
index, 57, 82, 101–102, 113, 115–116, 128, 139, 143, 144, 195
indirect, 70, 71, 99–100, 192
individual, 9, 10, 17, 18, 19, 59, 63, 75, 95–96, 98, 103f, 105, 125, 131–132, 141f, 142, 192
intensity, 17, 22, 55, 56t, 57, 58, 65, 66, 93, 99, 125, 141, 143, 146, 148, 185t, 195
limit, 59–62
mass, 16, 49, 58, 59, 95, 128, 131, 132, 133f, 192
method, 59, 64, 87, 90, 93–103, 105, 195
natural, 5, 29, 53, 62, 63, 193
pedigree, 59, 94, 95, 193
program, 96, 115
response, 13, 14, 15, 17, 18, 18f, 20, 21, 23, 58, 59, 60, 62, 82, 98, 103f, 128, 144, 145, 195
sib, 99
tandem, 57, 101, 196
truncation, 141, 185t, 196
Selective breeding, 6, 7, 8, 9, 10, 11, 13–23, 34, 59, 63, 64, 69, 72, 73, 88, 108, 119, 131, 144, 148, 151, 160, 164, 165, 167, 172, 177, 178, 179, 182, 183, 195
Self-fertilisation, 45, 46, 195
Selfing or self-fertilisation, 45, 46, 195
Sexual maturation, 15, 17, 18, 23, 74, 80, 97, 99, 115, 131, 137, 139, 147, 155t, 172, 177, 196
Shellfish, 1, 6, 7, 22, 31, 45, 46, 49, 52, 57, 59, 64, 67, 69, 75, 78, 84, 96, 102, 106, 109, 116, 126, 128, 146, 147, 148, 163, 165, 172, 178, 196
Sibs
full, 9, 33, 40, 41, 42f, 44, 45, 46t, 48, 49, 65, 72f, 73, 79f, 91, 95t, 96, 97, 98, 100, 101t, 102, 106, 108, 109, 113, 114–115, 123, 136, 140, 162, 190
half, 9, 36, 37, 44, 45, 46t, 52, 59, 65, 66, 91, 95t, 97, 98, 100, 101t, 102, 105, 108, 109, 113, 115, 117, 134, 136, 138, 158, 191
Spawning, 16f, 17, 23, 80, 99, 105–106, 111, 128, 133, 134, 155t, 163, 178, 196
Species, 2t, 6, 9, 10t, 15, 22, 22t, 27, 28, 29, 30, 32–33, 38t, 74, 88, 102, 120t, 121t, 140, 143–144, 154t, 155t, 173, 176, 196
Sperm, 11, 25, 26, 27f, 106, 134, 140, 166, 171, 173, 174, 196
Spermatogenese, 27f
Standard deviation, 22t, 32, 33–35, 55, 56t, 57, 84t, 114, 115, 116, 125, 137, 141, 185t
Strain, 14, 15f, 16, 17, 18, 33, 50, 51t, 52, 53, 65, 66, 88, 91, 120t, 121, 122f, 142, 144, 148, 166, 179t, 196
Stress, 6, 63, 67, 71, 75, 78, 80, 81, 143, 145, 147, 179, 180, 181

T
Tags, 70, 132, 136, 163, 176
Tank effect, 39, 109
Test station, 64, 121t, 137, 138–139, 196

Tetraploids, 169, 172, 173, 196
Triploids, 169, 171–173, 196
True breeding value, 91

U
Uniformity, 3, 48, 87, 180

V
Variance
  additive, 36
  effect of selection, 58
  environment, 109, 120, 123, 124
  genetic, 36, 37, 48, 50, 58, 87, 90, 98, 108, 109, 111, 148, 156, 160, 162, 191
  maternal, 109
Variation, 6, 9, 11, 15, 17, 18, 22, 23, 29–43, 49, 53, 54, 58, 60, 62, 65, 68, 69, 72, 73, 74, 75, 76, 77, 79, 81, 83, 84, 87, 95, 96, 98, 102, 103, 108, 109, 114, 119, 120, 121, 123, 124, 136, 138, 143, 148, 149, 152, 156, 160, 164, 167, 169, 171, 178, 179, 180, 196
Virus, 16, 20, 71, 72, 81, 146, 147, 196

X
XX females, 174

Y
YY male, 174, 197

Z
Zygote, 26, 166, 197

The manufacturer's authorised representative in the EU is Springer Nature Customer Service Centre GmbH, Europaplatz 3, 69115 Heidelberg, Germany. If you have any concerns regarding our products, please contact ProductSafety@springernature.com

Printed and bound by CPI Group (UK) Ltd, Croydon, CR0 4YY
07/01/2026
02030951-0003